Transboundary Migration in the Post-Soviet Space

Nikolai Genov / Tessa Savvidis (eds.)

Transboundary Migration in the Post-Soviet Space

Three Comparative Case Studies

PETER LANG
Frankfurt am Main · Berlin · Bern · Bruxelles · New York · Oxford · Wien

Bibliographic Information published by the Deutsche Nationalbibliothek
The Deutsche Nationalbibliothek lists this publication in the Deutsche Nationalbibliografie; detailed bibliographic data is available in the internet at http://dnb.d-nb.de.

ISBN 978-3-631-61485-3

www.peterlang.de

Contents

Preface

The post-Soviet geographical and political space offers nearly laboratory conditions for studying processes, which are global in their nature. This particularly concerns trans-boundary migration. Real boundaries did not exist on the territory of the former Soviet Union, only appearing after its dissolution in 1991. But still in most cases they are boundaries on political maps alone. This made the spontaneous trans-boundary migration on the post-Soviet space a mass phenomenon. In just one decade, the newly proclaimed sovereign Russian Federation had to accommodate a large wave of immigrants. The Federation became the second largest immigration country in the world following only the United States. The first massive group of migrants to the Federation consisted of ethnic Russians and Russophone communities. They hurried to leave the newly proclaimed independent republics under the pressure of nationalist movements and economic uncertainties. The next wave came mostly from Ukraine, Moldova, Armenia and Georgia. Currently the largest groups of migrants come to the Russian Federation from the former Soviet republics in Central Asia.

The individual motives for this large-scale migration have been and remain rather specific from case to case. But the most important driving factor for the largest groups of migrants was and continues to be the economic gap between the Russian Federation and all other post-Soviet republics, with the exception of the Baltic States. Economic migrants typically move for short or for longer periods to big cities in the Federation since they offer opportunities for regular or irregular labor. Especially during the nineties a substantial part of migrants' income was sent back to the countries of origin. In the literal sense of the word, these remittances made possible the survival of millions in the periphery of the post-Soviet space.

This largely concerned the remittances sent by hundreds of thousands of migrants from Armenia and Georgia who moved to the Russian Federation. The migrants wanted to escape from the local inter-ethnic, political and economic turbulences and to financially support their families in the difficult times. On the other side, the migrants filled in deficits in the workforce in some segments of the labor market in Russia's large cities. Millions of migrants settled there and contributed to the improvement of the otherwise catastrophic demographic development of the Russian Federation.

There are several studies focusing on the motives for out-migration from Armenia and Georgia, as well as on the various effects of out-migration from both countries. There are also studies on immigrants from the South Caucasus to the Russian Federation and particularly to Moscow. However, there has been no study so far simultaneously covering both sides of the migration chain. This could be surprising since one might expect full-scale descriptions and explanations of trans-boundary migration only in this way. The striking deficits of such studies are understandable. They would require substantial funding and well designed logistics in order to be implemented efficiently.

The generous funding provided by the Volkswagen Foundation to the research project *Out-migration from Armenia and Georgia* (ArGeMi, 2008-2010) made it possible to conduct this type of comparative research project. It directly relates out-migration from Armenia and Georgia to Moscow and the situation of migrants from Armenia and Georgia there. As attractive as it is, the project turned out to be rather challenging in terms of the development of a coherent conceptual scheme, the operationalization of key concepts, the elaboration of research tools, the control on field studies and the interpretation of primary information. In end effect, the invested work was very much rewarded by the possibility to consistently apply triangulation approaches in the analysis of the empirical findings and in their explanation.

We are very grateful to Volkswagen Foundation for this opportunity to carry out well coordinated comparative field studies in three highly interesting and dynamic environments (Armenia, Georgia and Moscow). In all three cases we had the fruitful cooperation of dedicated local teams from the Armenian Academy of Sciences headed by Prof. Dr. Gevorg Poghosyan, the Georgian Center for Population Research headed by Dr. Irina Badurashvili and the Russian State Social University in Moscow headed by Prof. Dr. Galina I. Osadchaya. We are grateful for their understanding and support.

Nikolai Genov and Tessa Savvidis

Nikolai Genov

Global City and Regional Periphery:

Challenges of Out-Migration from South Caucasus to Moscow

1 Introduction

With a registered population of ten and a half million Moscow is Europe's largest city. But in actuality, the population of the Russian Federation's capital is larger than the state statistics reports. Though the precise figures are unknown, a presumably substantial number of illegal (or irregular) migrants without a residence permit live in the city on a temporary or permanent basis (Grigor'ev and Osinnikov 2009). Both legal and illegal migrants originate from the provinces of the Federation and abroad. That said, the term 'migrant' is hardly ever applied to the numerous foreigners of Western European or North American origin who are employed by Russian and foreign firms, diplomatic offices or by international organizations in Moscow. In colloquial Russian usage, the term 'migrant' refers typically to a person who has come to Moscow from the former Soviet republics or certain other countries, such as China and Vietnam. This semantic issue is even more complex, as there are many ethnic Russians who came to Moscow from the former Soviet republics following the collapse of the Soviet Union. These individuals are not perceived as aliens, and are rarely referred to as 'migrants' when identified as newcomers to Moscow from Ukraine or Kazakhstan, for instance. The people typically called 'migrants' in Moscow originate from the Caucasus and Asia or at the very least look like they could come from those regions.

Whatever their country of origin, level of education, ethnic identification, religious affiliation or legal status, migrants are very much needed for the functioning of Moscow's economy. In one way or another migrants are being absorbed by the local division of labor, and production and services in Moscow heavily rely on their contribution. Without migrants from Tajikistan cleaning of the metropolis would collapse. Without migrant drivers from Ukraine the city's public transportation system would be at least partially paralyzed. Without migrants from Armenia and Georgia construction activities and the retail trade would face substantial difficulties. As seen from the opposite vantage point, the massive migration from the former Soviet republics now called the 'near abroad' was the major reason that the demographic balance in the Russian Federation was kept at least partly intact. Without these immigrants the decline of the Federation's population from 148 mln in 1991 to 142 mln in 2009 would have been much more significant. Nevertheless, some pessimistic projections tend to estimate that the population of the Federation will decline to around 100 million by 2050. In order to cope with this threatening demographic trend, Russian authorities recently introduced massive incentives to boost fertility. A second strategy was to make immigration attractive and easy, particularly for highly qualified immigrants. Due to a combination of these measures the population of the Russian Federation declined by only 50,000 in 2009, compared with an annual decline of 700,000 only several years prior (Zayončkovskaya, Mkrtčyan and Tyuryukanova 2009: 11 f.; Sovremennaya demografičeskaya situaciya 2010):

Figure 1: Net migration and population growth in the Russian Federation 1991-2009

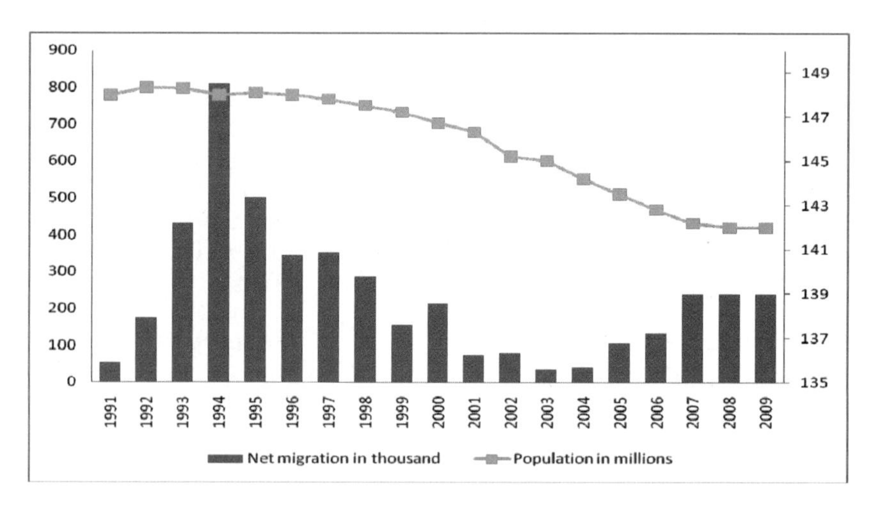

The attractiveness of the Russian labor market for citizens of the former Soviet republics is strong. In the years preceding the global economic crisis the economy of the Federation was booming due to high revenues from the export of crude oil and gas. The cleavage between the purchasing power of salaries and wages in Russia compared to that of the South Caucasus and Central Asia became substantial. This economic cleavage then turned into a powerful pull factor attracting millions of migrant workers from these parts of the former Soviet Union to Russia. The particularly strong growth of the construction sector was largely accomplished through the use of migrant labor, and the same can be said for the increasing number of jobs in municipal services. In addition, there were a number of intensive push factors that encouraged migrant workers to the Russian Federation. Due to the slow recovery of national economies and political turbulence in Central Asia and the South Caucasus, the remittances of labor migrants in the Russian Federation became crucial for the survival of migrants' families and the balancing of state budgets in both regions.

In spite of the Russian economy's great need for a large inflow of migrants, the autochthonous population across the Federation largely views migrants as an economic burden and cultural threat to Russian society. Yuriy Luzhkov, a former mayor of Moscow with 18 years in office, publicly expressed this belief in September 2010, generalizing that "migrants mean distortion of the economic system" (Migrants "distort economy" 2010). The perception that the ongoing shift of the Federation's ethnic composition is a threat to Russian society is widely shared. The slogan "Russia for the Russians" receives widespread support, and while there are particular traditional ethnic and religious groups specifically targeted by this slogan, migrants remain the general target. Luzhkov himself advocated stronger restrictive measures concerning foreign (cross-boundary) migrants to Moscow and the Federation.

The widespread and generally stable negative attitudes of locals towards migrants can be seen in numerous representative public opinion polls throughout recent years. Discussions regarding negative demographic trends and the needs of the economy to attract migrant labor have not changed this xenophobic attitude. It is well rooted in the Russian cultural and political tradition, and was ideologically and politically suppressed during Soviet times. Following the collapse of the Soviet Union these xenophobic attitudes remerged. Public opinion polls provide sobering information on the magnitude of this issue (Levada-centr 2009):

Table 1: "How do you relate to the fact that one may meet more and more frequently workers from Ukraine, Belarus, Moldova and other countries from the 'near abroad' on construction sites in Russia?" (N=1600, in percentage)

Relation	1997	2000	2002	2004	2005	2006	2007	2008	2009
Definitely and mostly in a positive way	25	26	30	21	22	20	21	18	19
In a neutral way	33	32	39	39	42	45	46	49	44
Definitely and mostly in a negative way	33	38	27	38	35	33	30	31	35
Difficult to say	9	4	4	2	1	2	3	2	2
Total	100	100	100	100	100	100	100	100	100

One might expect that the citizens of the global metropolis of Moscow would be more cosmopolitan and ethnically tolerant than the population of the Russian Federation on average. Data from public opinion polls of the Levada-centr do not support this notion. While in November 2009 just 18% of the national representative sample supported the slogan "Russia for the Russians", the share of the population of Moscow supporting it was 27% (Levada-centr 2009). Thus, xenophobic attitudes together with economic calculations determine the unfair or inhumane treatment of labor migrants, and particularly those on construction sites in Moscow.

All pressing issues of the necessary immigration and widespread negative attitudes towards migrants require political actions that are both well designed and well implemented. However, not unlike politicians in Western Europe, Russian politicians neglected the problems relating to alarming demographic trends and the mass inflow of immigrants for decades. It was only after a long delay that government on the Federal and local levels first attempted to develop and apply strategies for coping with the accumulated economic, political and cultural tensions due to mass immigration. Yet, the strategies for handling these tensions were inconsistent and remain inconsistent. Administrative measures vary widely between liberalization and restrictions. During the last decade one could register a shift of the political pendulum from a more liberal handling of immigration toward over-regulation with quotas and other restrictions. Such restrictive measures are usually inefficient, since strong elements of spontaneity are in the very nature of migration.

Given the great demand for a migrant labor force in the Russian Federation and particularly in Moscow, the negative attitudes of locals towards immigrants and the administrative efforts to restrict their inflow under the conditions of a demographic crisis are puzzling. This is provocative and relevant both from a scientific and practical point of view. In terms of social science research this puzzle cannot be solved by focusing only on the "demand side" of migration to Moscow, as has been attempted in the past. Likewise, a full-scale explanation of the processes cannot be achieved by focusing merely on the "supply side" of cross-boundary migration from the former Soviet republics. The aforementioned factors are so complex and interrelated that they require well focused multidimensional analyses. What motives really move so many migrants from the currently independent republics of the former Soviet Union to the Russian Federation and particularly to Moscow? How do they adapt to this social environment that is believed to be so unfriendly? How do migrants assess their stay in Moscow? Simple answers to these questions are not possible. Instead, the discussions must refer to broad social contexts.

2 Host Country and Donor Societies in Trans-Boundary Migration

The most recent available comparative data strengthen the point that the attractiveness of the Russian Federation for potential migrants is based largely on differences in the standard of living. This is exemplified by the substantial differences in the per capita GNI in the Russian Federation and most of the former Soviet republics (World Development Report 2010: 378-379):

Table 2: Gross National Income (GNI) per capita in selected republics of the former Soviet Union (2008, PPP $)

Country	GNI
Armenia	6,310
Azerbaijan	7,770
Belarus	12,150
Georgia	4,850
Kazakhstan	9,690
Kyrgyzstan	2,130
Moldova	3,210
Russian Federation	*15,630*
Tajikistan	1,860
Ukraine	7,210
Uzbekistan	2,666

The political conditions for the movement of labor from the "near abroad" to the Russian Federation were, and basically remain, favorable. There are no visa regimes between the member-states of the Commonwealth of the Independent States (CIS),[1] and the boundaries between them are porous. In some cases (Tajikistan, Armenia) the migration of labor force to the Russian Federation is specially facilitated by bilateral agreements. There are also special arrangements aiming to develop a free movement space of labor force between the Russian Federation, Belarus and Kazakhstan. After seventy years of co-existence under the umbrella of the Soviet Union there are many common value-normative grounds for mutual understanding and cooperation at work and leisure between citizens of the Russian Federation and migrants from CIS countries. Further, most migrants from these countries to Russia still have a good command of the Russian language.

Thus, there are a number of influential factors facilitating personal decisions to migrate from the South Caucasus or Central Asia to the Russian Federation in the search of employment and better life chances for one's family. The rational calculation of gains and losses takes into account visa regimes to most other desirable destinations, available information about living and working conditions in the Russian Federation, easy contact with networks of compatriots there, and the endurance of old business or human connections with citizens of Russia remaining from Soviet times. The adaptation to Russian conditions of life and work is in most cases not a new experience for migrants who still remember schooling in Russian language and culture.

Migration from the former Soviet republics to the Russian Federation is supported by a variety of intermediary economic actors such as labor agencies, transportation firms, firms offering legal services, etc. There are also well-

1 Diplomatic and other relations between the Russian Federation and Georgia changed abruptly following the war in South Ossetia in August 2008. Georgia left the CIS.

established civic and cultural organizations of the migrants' ethnic groups in major cities of the Federation, some of which date back to the Soviet era. Together with newly established ethnic organizations they support the adaptation of newcomers to the local circumstances in Russia.

Given these conditions the migration process appears to be profitable for all parties involved. The practically unused, or inefficiently used, labor force from the donor countries supplies the labor market of the Russian Federation where the demand for labor is high and will only increase in the future. The legal regulations of the migrants' stay are reasonably well developed, both in terms of the international treaties ratified by the Federation and in the Russian legal system. A special package of laws has facilitated the residential registration of foreigners, the legalization of their employment and their residency in Russia since January 2007.

In reality, the widespread xenophobic attitudes of the autochthonous population are only one part of the manifold problems accompanying the transboundary migration from the former Soviet republics to the Russian Federation. The depth of these problems is easily indicated by the seemingly simple matter of the reliability of statistical information on migration and migrants. Russian researchers report that there are striking discrepancies between the statistically registered migrant population and the estimated population of non-registered migrant workers. The differences between these two groups vary between 10 and 30 times for various donor countries from the former Soviet Union (Ivakhnyuk 2009: 33). This issue does not concern merely the precision of Russian statistics, nor inadequate legal regulations or inefficiency in law enforcement. Instead, behind this large discrepancy lurks the continuing problem of the widespread shadow economy in the Russian Federation. Both entrepreneurs and migrant workers are often inclined to agree to illegal labor relations without written contracts, disregarding basic legal regulations.

The motivations of migrant workers to move into the shadow economy are simple: some 70% of them send remittances to their countries of origin. The higher their salaries and wages, the larger the amount of remittances they can afford to send home. The reasons of entrepreneurs who choose to bypass legal work permission requirements for migrants are fundamentally related to the margin of profit. However, researchers are cautious to connect the simple relation of the profit calculation to the greed of entrepreneurs alone. Small and medium-size firms in Russia have notorious difficulties surviving in unstable local economic environments lacking transparency. Entrepreneurs are pressed by these circumstances to develop and apply coping strategies that often involve moving to the shadow economy. In some cases they are motivated to go in this direction not solely as a result of harsh competition, but also due to highly restrictive legal regulations. While these regulations should select a much-needed labor force and to allocate it where most needed, as is the case all over the world, these restrictive measures often strengthen the temptation for businesses to move into the shadow economy. This occurs with particular frequency in times of economic crisis. These factors have led researchers to assume that only one third of the migrant labor force in the Russian Federation is currently employed outside of the shadow economy. Not surprisingly, there are then a number of implications generating from this situation. First, lower wages for the irregular migrant labor force undermines competition in the labor market. Further, the working conditions of the irregular migrants are often worse than standard working conditions in the Russian Federation. According to estimations, irregular migrant workers receive nearly the same salaries and wages as local workers, but migrants are often expected to work extremely long hours, ranging from 60 to even 70 hours per week without days off, and often under particularly bad conditions (Vishnevskiy et al. 2010).

The illegal employment of migrant workers is most typical in the Russian construction industry, where migrants likely constitute the majority of those employed. Activists from the international organization *Human Rights Watch* recently documented grave violations of laws and regulations concerning the human rights of migrant workers in the industry. The infringement on legally guaranteed human rights starts as early as the promotion and advertisement of employment on Russian construction sites for migrants. Both Russian and local agencies in the CIS states advertise the positive work conditions on construction sites, the majority of which do not actually exist. Next, transportation firms and other intermediaries commonly cheat the labor migrants in their passage to these construction sites. Contrary to the clear legal regulations, written labor contracts are rarely signed between the construction firms and migrant workers. Moreover, the contracts that do exist rarely include precise definitions of the conditions of labor. This lack of clarity is presumably intentional for the purpose of tax evasion, or the evasion of liabilities for employers in the case of work-related accidents. Working conditions are often inhumane in terms of long hours, dangerous working environments and low salaries. In addition, the payment of salaries and wages is typically delayed, not paid in full, or in some cases not paid at all. Likewise, the abusive treatment of labor migrants by supervisors has been documented in numerous cases. The state institutions responsible for the protection of migrant rights usually act slowly, if they intervene in these cases at all. Moreover, representatives of the local administration, legal institutions and police have been known to further abuse the unclear or illegal status of labor migrants through the application of various forms of extortion. The options for effective action, complaints or resistance of exploited, abused, or even tortured migrant workers are greatly limited (Buchanan 2009).

It would be naïve to assume that the illegal employment of migrants and the ensuing cheating and exploitation are unique to the Russian Federation. In fact, this happens to migrant workers on a daily basis in all corners of the globe. The commoditization of the labor force of migrants everywhere follows the market rules of pursuing maximum profit, and not necessarily maximum humanism. However, the institutional conditions in the Russian Federation still allow extremes of wild capitalism, which does not apply to the treatment of migrant workers alone. In many cases both workers of ethnic Russian origin and migrants are similarly badly treated. But workers of foreign ethnic origin are particularly vulnerable since they are also subject to xenophobic attitudes and actions, and often cannot rely on any legal protection due to their illegal status.

Closely related to this is another issue that is becoming increasingly relevant for the study and management of trans-boundary migration to the Russian Federation and particularly to Moscow. The large wave of repatriation of ethnic Russians and russified segments of other ethnic groups to the Russian Federation has past. The potential of labor migration from Ukraine, Moldova and Belarus is declining, and the same applies for countries in the South Caucasus. Therefore, the former Soviet republics in Central Asia increasingly become the Federation's most relevant source for migrant labor, due in large part to the low level of economic development, high unemployment and high fertility rates in these countries. The acceleration of the shift in the ethnic composition of migrants has become obvious in recent years (Vishnevskiy et al. 2010):

Figure 2: Changes in the ethnic composition of labor migrants to the Russian Federation (2006-2008, as % of all labor migrants)

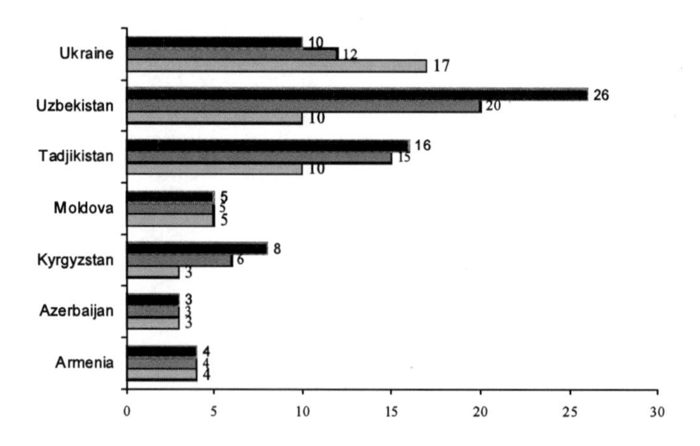

The change in the ethnic composition of the migrant inflow to the Russian Federation is accompanied by numerous problems, and has become one of the Federation's major domestic challenges. The newcomers from Central Asia typically have low educational levels, and thus can take only low-skilled positions in the division of labor while the Federation is in urgent need of highly skilled workers as well. For instance, following the large-scale emigration of well-educated individuals of Jewish and German ethnic origin, one fifth of the employment positions in Russia's health care system are currently vacant. The need for specialists is becoming more and more crucial, with the replacement of the current labor force depending on a shrinking number of younger cohorts. The command of the Russian language by younger immigrants from Central Asia is typically at low levels, if present at all, and their economic, political and cultural integration into the multicultural but predominantly Russian society is the subject of intensive concerns.

These problems have great relevance in economic, political, social and cultural terms, but their solutions cannot be achieved simply through the introduction of quotas for immigrants from particular countries or those with a specific occupational profile. Further, these problems do not concern the access of migrants to the Russian labor market alone, but instead a more general issue relates to their inclusion in all activities of the host society. This is not a specific problem to Russian society. The many facets of the problem have been subject for serious concern in all Western European societies for decades. One should note, however, that the features of the problem challenging the Russian society are more complicated than the similar problems of integration of migrants in Western European societies. One of the major reasons for this difference is related to the fact that Russia's autochthonous population is presently comprised of one-fifth Muslims, with a higher fertility rate and a longer male life expectancy than that of the Slavic population.

In this broad context the special focus of the following analyses and conclusions will concern only the moving forces, processes and effects of immigration to Moscow. It is the only city from the Russian Federation and the former Eastern Europe that is regarded as truly global in terms of business activities, human capital, information exchange, cultural experience and political engagement. As measured by these criteria, Moscow ranked 25th in the world scale of 65 global cities in 2010.[2] The city is by far the leading urban center in the Russian Federation. Though Moscow comprises roughly just 7% of the Federation's population, it produces one quarter of its GDP, giving Moscow an economic performance equivalent to that of South Africa. More than a third of the foreign direct investments in the Russian Federation are made in Moscow. The city is particularly attractive for migrants since the per capita income in Moscow is three times higher than the average per capita income in the Federation. Thus, Moscow is both a key component of the global inter-city networks and commodity chains (Sassen 2010).

In late 2008 a new master plan of the city was adopted, and is set to be in place by 2025. The plan lays stress on the construction sector – the gentrification of old buildings, construction of housing and offices, development of the trans-

2 The ranking is led by New York, Berlin is in the 16th and Munich is in the 33rd position (*The Urban Elite* 2010: 3).

portation and tourist infrastructure, substantial improvements of the water supply and sewer systems. These ambitious plans can only be materialized through the intensive and efficient use of the migrant labor force. However, one should keep in mind that the very concept of "migrant labor force" is somewhat opaque as it concerns Moscow. Half of the city's registered residents were not born there, so in this sense Moscow is a city of migrants. Migrants were the major source of the recent increase in the city's registered population as reported by the state statistical office. Various estimations add to the legal immigrants with residence registration (propiska) around 1.5 millions migrants, mostly from CIS countries, which permanently reside and work in Moscow without work permission and typically without residence registration. Large parts of them are illegally employed exactly in the construction industry. Changes of legislation urged many of these migrants to legalize their status and thus, contribute to the statistically registered population of the capital city. Still, the estimated number of illegal migrants in Moscow remains high (Zayončkovskaya 2009).

In order to facilitate the inflow of skilled migrants, the Moscow municipal government instituted a program in 2008 that stresses the development of a precise information system concerning the real migrant labor force needs of the city's economy. Based on this information, the control of immigration flows should be improved in order to reach a more efficient match between the needs of Moscow's economy and the parameters of the migrant inflow. In addition, the administrative measures aiming at the efficient integration of immigrants should be better targeted, designed and financed. Following these decisions, a new code of conduct for immigrants is being developed. Surprisingly or not, its major rules for immigrants include a prohibition of killing animals in public and wearing clothes or dress specific to one's country of origin, and a requirement to speak Russian.

The above rules are, first of all, obviously intended to apply to immigrants from Central Asia to Moscow. The research project *Comparing Out-migration from Armenia and Georgia* and its special focus on Moscow as destination for migration dealt with two groups of migrants, which do not, as a rule, cause the problems mentioned above. There were other more specific reasons to look at these two groups of migrants to Moscow as the focus of this study. Specifically, the out-migration from Armenia and Georgia was selected for the comparative study because of both the striking similarities and differences of these cases. Both Armenia and Georgia experienced a sharp decline of their respective GDPs during the nineties, and in both societies economic recession was accompanied by interethnic wars. In each of the countries large groups of individuals were encouraged to emigrate in the search for income, security and better future for their children. As a result, the populations of Armenia and Georgia each decreased by one million due to mass out-migration after 1990. The largest group of these migrants headed to the Russian Federation and to Moscow in particular, either on a permanent or temporary basis. In both cases, migrants from Armenia and Georgia regularly sent remittances to relatives in their countries of origin, as do most migrants from post-Soviet societies.

These similarities notwithstanding, the conditions for out-migration are still significantly different in Armenia and Georgia. Armenian society is at present nearly homogenous in ethnic terms. Russians and representatives of other ethnic groups left the country immediately after it proclaimed its independence in 1991. The out-migrants thereafter have been ethnic Armenians who typically migrated due to economic motives. Georgian society, on the other hand, is multiethnic. Out-migrants from Georgia might be ethnic Georgians, or perhaps ethnic Armenians, Ossetians, Abkhazians or representatives of other ethnic groups. Some of these migrants are refugees or might be motivated to migrate due to ethnic considerations. This makes the study of migrants from Georgia significantly more complicated than the study of migrants from Armenia. In order to unify the approach the research team decided to study "migrants from Armenia" and "migrants from Georgia" and not "Armenian migrants" or "Georgian migrants" in Moscow.

The above differences have some foundation in the relationships between the states of origin and the state of destination. The relationships between independent Armenia and Russia have been traditionally friendly, and there are agreements in place between the two regarding labor migration. On the other hand, independent Georgia and the Russian Federation have had a difficult relationship in the last decade and waged war in August 2008. Hence, one could expect different patterns of adaptation of migrants from Armenia and Georgia in Moscow, as well as differing attitudes of the population and authorities in Moscow towards migrants from both countries. These assumptions could be tentatively tested by comparing primary data obtained in interviews carried out in the spring of 2009. In the following paragraph only information obtained from migrants who have been employed (including self-employment and business activities) in Moscow will be presented and interpreted.

3 Labor as a Commodity on the Regional Market

Trans-boundary migrants are involved in processes and relations, which intensively challenge social theory and policies. For the social sciences, the main issue is rooted in the complexity of the out-migration and immigration processes, which "create whole new ways of linking labor-exporting and labor-importing countries" (Sassen 2007: 141). The crucial point in this practical and theoretical complexity is the understanding of labor as a specific commodity. Its circulation follows the general rules of the dynamic relationship between supply and demand, but also has substantial specific components compared with other subjects of market exchange.

Nowadays the interpretation of labor as a commodity sounds like a tired cliché. However, it was not long ago that ideology precluded any market-oriented understanding of labor and labor migration in Eastern Europe. In fact, the phrase "labor is not a commodity" was a cornerstone in politics and the social sciences in the region prior to 1989. The background of this sentiment was the understanding of labor as commodity under capitalism, which was first introduced in the early writings of Marx (Marx 1970 [1847]: Ch. 2). Under the conditions of socialism the situation was supposed to be radically different, and labor was assumed to be free of commoditization since the means of production were widely socialized (state-owned). Economy was thus dominated by plan and not by market, as there was officially no labor market and labor had no market value. Stalin put the point bluntly, concluding scientific discussions of Soviet economists in his typical style: "Talk of labor power being a commodity, and of "hiring" of workers sounds rather absurd now, under our system: as though the working class, which possesses the means of production, hires itself and sells its labor power to itself" (Stalin 1952: 17).

At first glance, this statement and its substantiations were correct. Under state socialism wages and salaries were indeed determined by administrative decisions and not by the market equilibrium of demand and supply of labor. However, a closer look at the realities of state socialist societies reveals a controversial picture. Millions of peasants moved voluntarily from rural areas to the emerging industrial agglomerations because of one major reason – the remuneration of industrial work was higher than the remuneration of work in the agricultural cooperatives or state-owned farms. This process can be clearly defined in market terms: peasants moved to towns because they wanted to receive better pay for selling their labor force in industrial enterprises.

Following the first wave of extensive industrialization, deficits in the labor force in most Eastern European societies became chronic, resulting in the large-scale demoralization. Directors of industrial enterprises tried to cope with the deficiencies of socialist labor organization and the lack of motivation for work using counterproductive means, which further worsened the situation. They tended to overstaff their factories by offering benefits to attract workers from other factories. Since salaries and wages were generally kept low by administrative means there was little room for bargaining over remunerations. But directors could offer employees other types of benefits, such as housing, kindergartens, vacation homes, on-site medical services, a lower pension age, etc. Thus, workers would be best positioned to sell their labor force under better terms of trade, which fluctuated greatly from factory to factory.[3] In addition, there was an administratively regulated international labor market amongst socialist societies. Thousands of Vietnamese and Cuban workers voluntarily went to work in Eastern Europe for better pay. Thus, the reality of the labor relations was inherently contradictory under state socialism, and did not function purely according to

3 During the eighties, annual labor fluctuations of 10-15 percent of factory staff were perceived as normal in Eastern Europe. No administrative measures could stop the process of selling the labor force under better conditions.

ideological and political assumptions. On the contrary, state socialist societies were increasingly moving away from the idealist notion that labor could be free of commoditization. It became clear that the slogan "from each according to his ability, to each according to his needs" could not be practically achieved. The ideologically guided efforts to develop a convincing alternative to the efficiency of motivation, discipline and allocation of labor force under market conditions could not last for long.

The rigid administrative regulations of the Eastern European labor markets dissolved along with the political changes after 1989. The region's population reacted quickly. In only a few days, thousands and later millions of Eastern Europeans joined the flows of international migrants offering their labor in markets all over the world. No protectionist barriers in Western Europe and North America could stop them. When struggling with a plethora of difficulties in their adaptation to the economy, politics and culture in host societies, Eastern European migrants experienced, and continue to experience, the shadowy sides of globalization. One of these directly concerns the labor market, which is much less globalized than the markets of goods, services and finances.

In spite of all the challenges regarding trans-boundary migration, twenty years after the beginning of the transformations in former Eastern Europe, its effects on the region are spectacular. The lifting of administrative restrictions for the international movement of labor resulted in a net loss of one-third of the population in Armenia and Georgia. But Armenia and Georgia are not unique in this respect. Millions of migrants left Poland, Romania, Bulgaria, Moldova and Ukraine on a temporary or permanent basis. Some 80 percent of these migrants are assumed to have joined international migration in the search for employment. The economically motivated out-migration from Eastern European societies continues and will continue in the foreseeable future. Its most visible and long lasting effect is the net loss of population, the state of which was not improved by the enlargement of the European Union to Eastern Europe. Several of the new member states of the European Union are particularly affected by mass out-migration (Europe in Figures 2008: 75):

Table 3: Net migration projections for the Eastern European members of the EU (1000)

Country	2006-10	2011-15	2016-20
Bulgaria	-58.3	-77.1	-83.1
Czech Republic	16.9	-10.2	22
Estonia	-9.9	-13.5	-7.2
Hungary	69.3	33.1	52.2
Latvia	-12.6	-22.1	-11.7
Lithuania	-29.5	-35.1	-18.7
Poland	-158.7	-277.3	-153.7
Romania	-68.4	-168.6	-226.5
Slovakia	-11.9	-13.7	-2.8
Slovenia	30.1	18.7	22.8

The above migration processes were determined by multiple interconnected variables and have fluctuated substantially due to a variety of factors, which is precisely why it is difficult to specify lines of determination and their effects. Only tentatively could some elements of trans-boundary migration be identified in relative isolation and incorporated in explanatory models. An all-encompassing explanatory scheme concerning the phenomenon of migration in general or trans-boundary labor migration in particular does not exist. This will likely continue in the near future, despite the fact that the classical "push-and-pull" conceptual scheme has long been successfully used in descriptions and explanations. However, if an analysis and search for explanations were to go sufficiently deep, one would find that there are numerous push and pull factors interacting at various structural levels with varying intensity and effects. Other conceptual schemes, either more or less elaborated, should be used to explain this range of specific push and pull processes. Some of the explanatory schemes focusing on specific conditions of the emigration / immigration of labor could be briefly and selectively summarized as follows:

Macro-social conditions of trans-boundary migration

Macro-economic conditions

There are very substantial cleavages between labor productivity, GDP per capita, level of wages and salaries and standard of living amongst societies across the globe. These cleavages attract labor from countries with surplus labor force and lower income levels to countries with labor force deficits and higher income levels, following the rules of balancing demand and supply. The issue was exemplified by the differences in GNI per capita in selected post-Soviet republics. In addition, labor markets in societies hosting migrant labor are always segmented. This segmentation facilitates the attraction of the migrant labor force and its allocation, mostly in segments marked by the worst labor conditions and the lowest remuneration.

Macro-political conditions

Host societies differ substantially in their political visions and legal regulations regarding immigration in general, and particularly concerning the immigration of labor force. Some advanced societies attempt to manage their deficits of labor through technological and economic restructuring and measures of qualification and re-qualification of the available human resources. Others rely on well-designed and carefully implemented immigration policies focusing on attracting those labor resources most needed in their economies. These different policies substantially modify the influence of push and pull factors for migration of the labor force. The same applies for the policies of the 'donor' states, which can prohibit or support out-migration. Since the 'donor' states are often rather weak they typically do not intervene in the process. Instead, modifications of push and pull factors are being increasingly determined by the decisions and policies of supranational integration bodies, like the European Union, the Commonwealth of Independent States (CIS) or NAFTA. All these intervening variables should be taken into account in conceptual schemes aiming at systematic description and explanation of the migration of the labor force.

Macro-cultural conditions

The cultural proximity or cultural distance of the donor and host societies in trans-boundary migration might become substantial factors fostering or hindering the out-migration and integration of immigrants. The level of proficiency in the language of the host society might be a particularly relevant factor for the decision to migrate. This same factor might also be significant for the success or failure of immigrants to integrate into the economic, political and cultural structures of the host society.

Meso-social conditions of trans-boundary migration

Meso-economic conditions

There are many different types of economic actors facilitating or hindering trans-boundary migration. Some of these actors make legitimate profit by offering services to migrants, while others exploit or abuse them. The range of actors involved at the meso-level of economic transactions includes employment placement and travel agencies, firms offering translation and legalization of documents, and so on. There are also various criminal or semi-criminal networks offering document forging services, trafficking, mediation in the search for irregular jobs, irregular legalization of residency, etc.

Meso-political conditions

Both in the countries of origin and host countries exist various political formations (parties, movements, politically-biased non-governmental organizations, informal groups, etc.) at the local level whose activities either foster or hinder out-migration. The suppression or persecution of ethnic and religious minorities or political opponents might become a powerful push factor, motivating victims to go abroad in search of income and security. Alternatively, constructive activities of local political organizations or NGOs in the host countries may substantially improve the chances for successful economic, political and cultural integration of immigrants. On the contrary, the xenophobic activities of political parties, non-governmental organizations, and nationalistic groups or skin-heads in host countries can greatly hinder the integration of immigrants into local economic, political and cultural structures.

Meso-cultural conditions

The lack of high quality educational establishments in the locality of origin of potential out-migrants is often a serious argument for them to take the decision to migrate. In many cases concerns about the quality of education are not focused on the personal educational needs of the potential out-migrants themselves, but instead on the educational needs of their children. As seen from another vantage point, one of the crucial conditions for integration into the host society is command of the local language. Thus, the very availability of institutional options for learning the language of the host society, as well as the economic affordability of learning in educational establishments, might become decisive for the quality of the economic, political and cultural integration of immigrants.

Micro-social conditions of trans-boundary migration

Micro-economic conditions

In simplified interpretations of trans-boundary labor migration the individual migrant is often presented as a pure *homo oeconomicus,* primarily calculating economic advantages and disadvantages of leaving one's home country temporarily or permanently. The crucial factor for this decision is the possibility to earn a higher income in the host country, and thus afford a higher standard of living and a better quality of life. No doubt, in many cases this interpretation is valid and can be easily operationalized in the study of well-defined push and pull factors. The next step is their interpretation as variables in a 'rational choice' decision. However, this is a rather superficial interpretation for those cases in which out-migration is practically the only option for securing the survival of the potential migrant and his or her family.

Micro-political conditions

Potential migrants do not merely egoistically calculate the economic advantages and disadvantages of out-migration for themselves, but being *zoon politikon* they usually consider the advantages and disadvantages of out-migration also in altruistic terms, particularly when considering the futures of their children or remittances to relatives and friends. This balancing of personal interests and social responsibilities in individual decision-making and behavior might also include broader social considerations regarding the development of the locality or the country of origin. A specific dimension of the migration process in this context is the development and use of interpersonal informal networks based on common ethnic origin, religious affiliation, clan and family belonging, origin from the same locality, and so on. These informal "weak ties" in a micro-polity sometimes play a crucial role in the decision to migrate, as well as in overcoming the constraints related to migration in conditions of uncertainty and lacking transparency. Conditions of this type are particularly relevant for individual decisions in times of rapid societal transformations.

Micro-cultural conditions

Value-normative orientations developed through the socialization process might facilitate or hinder the decision of an individual to migrate from the country of origin or to integrate in the host society. Some of these orientations can have ethnic or religious characteristics. Value-normative orientations focusing on professional development and realization take on a growing relevance in personal decisions and behavior related to trans-boundary migration. Patriotic feelings and related cultural orientations might become influential factors in the decision making of individuals regarding migration.

Against this background of complex conditions determining the decisions and behavior of potential or actual migrants, it becomes clear that efforts to develop and apply explanatory models must cope with multiple contingencies. One may focus on the concept of cultural preferences or rational choice among alternatives in the attempt to explain the relative importance of push and pull factors influencing personal decisions to leave one's home country or to choose a specific country of destination. The decision itself might come about to a great extent due to the search for employment or better pay, economic and political security, better professional development and realization, a better future for the children, etc. As seen from a macro-social perspective, when searching for explanations one may focus on the segmentation of national labor markets or on center-periphery relationships in economic, political and cultural terms. In addition, serious theoretical and empirical research is possible and necessary in order to determine and explain the implications of emigration for the societies of origin, as well as for host societies. For instance, out-migration reduces economic and political tensions and conflicts due to mass unemployment in migrants' societies of origin, but at the same time usually deprives these societies of their most active labor force.

From another vantage point, one must take into account the relevance of the remittances of labor migrants for the survival or wellbeing of their families in their home countries, as well as for the stabilization and development of national economies. As seen from yet another angle, one must also include the probability of return of labor migrants to their countries of origin in the explanatory model. Migrants might return with accumulated financial resources, work experience, cultural enrichment, established social networks, etc. A special research field with policy relevance concerns the status and behavioral patterns of labor migrants in their host countries. The task to develop precise descriptions and explanations of the immigration process is, in and of itself, multidimensional enough. The labor markets in host countries might have large niches for newcomers or could be saturated with all types of labor supply. The population of the host country might be largely tolerant to newcomers, but it is also possible that deeply rooted hostility toward all foreigners or toward foreigners from specific countries, regions, ethnic belonging or religious affiliation dominates. Negative attitudes regarding labor migrants might be prevalent even when all rational arguments support the point that migrants take jobs which locals do not want, that immigrants substantially contribute to the GDP growth and typically do not abuse the social security system of the host society.

This large variety of approaches, interpretations, cognitive results and policy orientations reflects the many facets of trans-boundary migration, which is being conditioned by numerous factors and has effects at various structural levels of society. Focusing mostly on the orientations, decisions and actions of individuals, this complexity might be schematically reduced in the following way[1]:

1 Schematic presentation by the author

Figure 3: Moving factors, processes and effects of trans-boundary migration

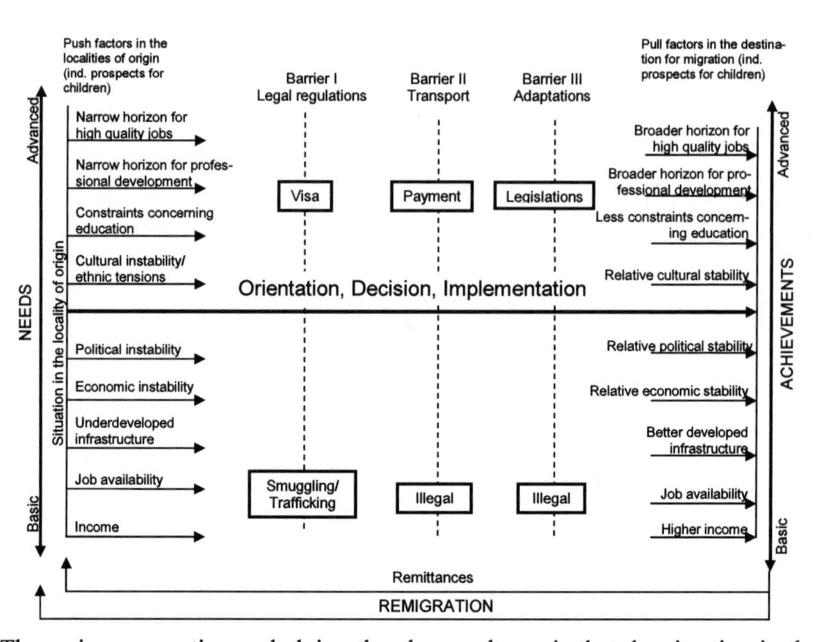

The major assumption underlying the above scheme is that the situation in the country of origin *and* in the host country of migration as well as orientations *and* actions, desires *and* practical experience should all be taken into account simultaneously. The conceptual background for resolving this issue developed from the pyramid of needs (Abraham Maslow) and from classical ideas regarding the causes, processes and effects of social mobility (Pitirim Sorokin). These classical ideas, as well as recent developments in social science theorizing and research, were incorporated in the conceptual scheme. The links and processes are determined by multiple interconnected factors and fluctuate substantially from situation to situation and over the course of time. Only tentatively could some elements of trans-boundary migration be identified in relative isolation and might be well embedded in explanatory models. As complex as they are, these models do not represent an all-encompassing explanatory scheme on the variety of causes, reasons, processes and effects of trans-boundary labor migration. As indicated above, a conceptual scheme of this complexity does not exist as of yet. However, there are some relatively well-elaborated partial conceptual schemes, which might be used for explaining specific push and pull processes and their effects. In the ArGeMi project the descriptive and explanatory processes followed the economic, political and cultural lines of determination at the level of societies (macro-social level), local institutions, organizations, movements (meso-social level) and the level of individuals and small groups (micro-social level).

The ideas presented in this way guided the operationalizations of concepts and the elaboration of research tools. The major part of this was focused on the perceptions, orientations, decisions, actions and experiences of realized or would-be trans-boundary migrants. A questionnaire for face-to-face interviews with migrants who had returned from Moscow was applied on 200 individual cases in Armenia and on 200 cases in Georgia. Another questionnaire was prepared and applied for face-to-face interviews with 300 returnees to Armenia and 300 to Georgia from other countries than the Russian Federation. A third questionnaire was designed and applied for face-to-face interviews with 100 individuals in both Armenia and Georgia who had no migration experience but had the intention to go abroad. The interviews were conducted in five towns in Armenia and five in Georgia. In order to enrich the informational background for the interpretation of the findings obtained in the interviews, two additional tools were designed and applied in both countries. The first was a questionnaire for structured expert interviews. The second was a tool for the structured monitoring of events, including both descriptive and analytical components. In Moscow a questionnaire for face-to-face interviews with 200 migrants from Armenia and 200 from Georgia was designed and applied, under the condition that only those migrants who came to Moscow after 1990 would be interviewed. A questionnaire for structured interviews with experts in the field of immigration studies and policies in Moscow was also designed and applied, along with a tool for the structured monitoring of events. The field studies in all three countries were implemented in two rounds, in the spring of 2009 and the spring of 2010.

The crucial objective in the preparation of the research tools and field studies was to achieve triangulation in the explanation of relevant empirical findings. This was achieved through the composition of tools containing built-in controls, as well as the mutual control of information obtained by the application of several research tools. Since the field studies covered migrants from Armenia and Georgia currently in Moscow, as well as returnees from Moscow interviewed in Armenia and Georgia, the findings allow comparisons both in a synchronic and diachronic time perspective.

4 Hypotheses, Empirical Findings and Interpretations

Numerous specific findings are the subject of detailed discussions in the following chapters. The aim of the present paragraph is only to focus on one limited area of hypotheses, findings and interpretations. This is the area of out-migration from Armenia and Georgia to Moscow in search of employment (also self-employment and other business activities). This specific out-migration concerns the largest segment of migrants from Armenia and Georgia to Moscow. Moreover, findings regarding this type of migration are particularly indicative for the general trends in out-migration from Armenia and Georgia. The relating analysis and conclusions are guided by the assumption that following the fully spontaneous migration processes during the nineties it is now a pressing task to reduce uncertainties in the orientations, decisions and actions of labor migrants, as well as in the steering of the migration process by state administrations. For this limited purpose data from the interviews conducted in Moscow with migrants from Armenia ("Current migrants from Armenia in Moscow") and migrants from Georgia ("Current migrants from Georgia in Moscow") will be briefly analyzed. The aim is the synchronic comparison of the adaptation of labor migrants from both countries at the time of the field studies (spring 2009) in Moscow. In order to introduce a diachronic comparison, data from the interviews conducted in Armenia with returnees with labor experience in Moscow ("Former migrants from Armenia in Moscow") will be introduced as well. The same will be done with data collected in interviews in Georgia with returnees also having labor experience from Moscow ("Former migrants from Georgia in Moscow"). The data on labor migrants (employed, self-employed, or involved in business activities) will be separated from the broader samples and treated as sub-samples of 100%.

One key hypothesis on labor migrants' coping with uncertainties in Moscow refers to the strength of the weak social ties (informal interpersonal relationships). There are several reasons for this special focus. The major reason is the repeated insistence of the interviewed experts in Moscow and in both countries of origin that the information and practical support provided by compatriots, friends and relatives have been and continue to be the most important form of social capital for the migrants. Thus, the informal "weak ties" are persistently conceived as the most decisive factor influencing the first and most difficult task facing labor migrants in Moscow – finding employment. This assumption of the experts seems plausible, but its verification was successful only in part. It turns out that the relative relevance of informal support has substantially changed over the years. Currently, the social capital of personal relations is no longer so decisively relevant in resolving the issue of employment, as has been assumed by experts. Moreover, the experts believed that the larger Armenian Diaspora in Moscow would be more supportive to the newcomers from Armenia than the smaller and less influential Georgian Diaspora to the newcomers from Georgia. The data both partly supports and partly challenges this second hypothesis:

Figure 4: How did you arrange your employment/self-employment in Moscow?

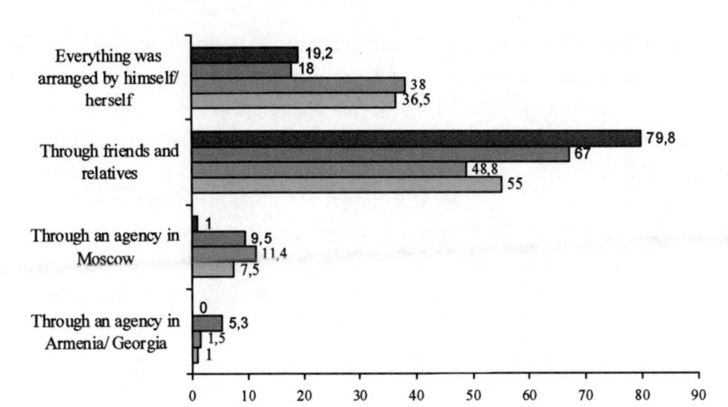

No doubt, the "weak ties" of the informal relations of labor migrants with friends and relatives have been particularly important and supportive for the first wave of labor migrants from Armenia and Georgia who came to Moscow following the collapse of the Soviet Union. The interviews with currently employed migrants from Armenia and Georgia in Moscow show a different picture. The relevance of informal interpersonal relations has declined substantially, although these "weak ties" still remain the strongest element of support for the newcomers. But the relevance of factors supporting the search for a job has shifted towards the reliance on personal initiative and capacities to get properly oriented and make adequate decisions. This is an indication that the new migrants are now better informed than those in the past, and that the labor market in Moscow has become more transparent.

The experts' second assumption was falsified by the empirical findings. Contrary to what they expected, the migrants from Georgia profited most from the support of friends and relatives. This was the case in 79.8% of the interviewed labor migrants interviewed in Georgia who were previously employed in Moscow. Support from informal relationships was a deciding factor to a smaller proportion of labor migrants with previous experience in Moscow interviewed in Armenia (67% of the cases). This finding is surprising only at the first glance, as the explanation is actually quite simple. Because Georgians were less active in migrating before the dissolution of the Soviet Union, labor migrants from Georgia to Moscow immediately required more intensive support by their compatriots, friends and relatives and relied on this support. Therefore, the data do not support the frequently expressed belief that the relatively large community of people with Armenian ethnic origin would more actively and efficiently help the new migrants from Armenia to Moscow. An explanation of this must take into account the specifics of the old and new Armenian Diaspora. Although there are estimated half a million regular and irregular residents of Moscow of Armenian ethnic origin, they belong to seven rather different groups, and people belonging to these different groups usually do not communicate with one another (Galkina 2006: 186-187).

Thus, it is indicative for the general normalization of the labor market for migrants in Moscow that the current labor migrants from Armenia and Georgia manage to arrange their employment in Moscow alone in the same proportion. One might then question why migrants so seldom rely on formal organizations (and particularly labor agencies) for arranging their employment. The explanation of this phenomenon should make reference to a variety of circumstances. First, most agencies in Armenia and Georgia designed to support labor migration, and particularly to arrange work places for migrants in Moscow, were discovered to have strong criminal links. Consequently, their licenses were withdrawn by the authorities in both Armenia and Georgia, thus leaving a void of institutionalized mediators between job seekers in Armenia and Georgia and the labor market in Moscow. This type of agency has only just recently reappeared in Armenia, but the impact and efficiency of their involvement in resolving the major task of labor migrants is still greatly limited. The support of employment mediation agencies has been called for in Moscow, but can be related to just a small share of migrants from Armenia and Georgia. This is a clear indication of the immaturity of these services. For the most part, they are still regarded as unreliable and expensive. Moreover, the worsening of the relationship between the Russian Federation and Georgia had the consequence that migrants from Georgia could only in very special cases utilize the services of these agencies. This is a notable case, deviating from the picture of normalization of the labor market and development of related services in Moscow towards maturity and transparency.

The disparity between the hypotheses of the experts and the data obtained in the field studies also applies to the employment of migrants in various sectors of Moscow's economy. Both in Moscow and in Yerevan the experts were nearly unanimous that Armenian labor migrants in Moscow have traditionally been and continue to be predominantly employed in Moscow's construction industry. As for the labor migrants from Georgia in Moscow, both experts in Moscow and in Tbilisi were fairly certain that they have been and currently largely involved in the retail trade, and in activities at the fringe of legality or even beyond its boundaries. The data obtained from the interviews conducted in the spring 2009 shows a different picture both concerning the type of employment of former and the current labor migrants from Armenia and Georgia in Moscow:

Figure 5: In which sector of Moscow's economy you have been (you are) employed?

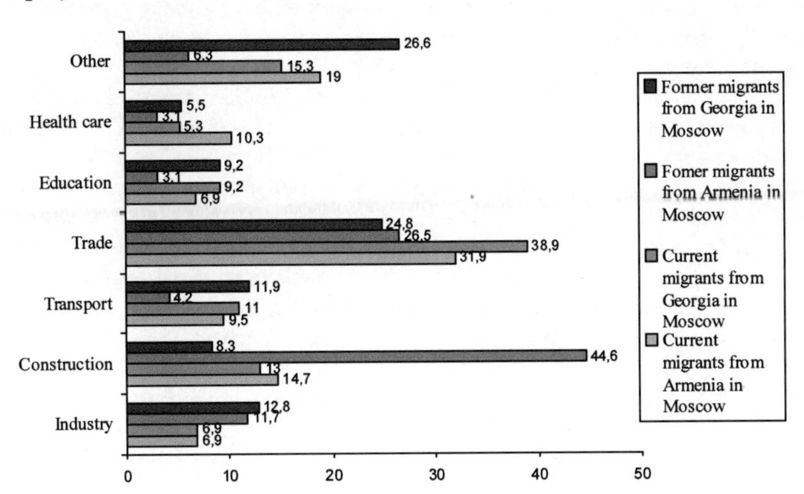

The assumption about the predominant employment of Armenian migrants in Moscow's construction industry turned out to be not completely incorrect. The assumption of the experts concerning the employment of Georgian migrants in Moscow was also not entirely false. But both opinions were based on outdated information. Armenian migrants from the first large wave of labor migration typically did make their living by working on Moscow's construction sites. And migrants from Georgia often did make their living involved in so-called 'other activities'. But the situation has since changed. The reported current employment of migrants from Armenia and Georgia reveals well-defined niches for typical labor activities of migrants in Moscow. The previously striking differences between the sectors of employment of migrants from both countries have disappeared. Currently migrants from Armenia and Georgia are employed in nearly the same proportions in Moscow's industry, construction business, transportation, education and the so called "other" activities, which usually include various services. The relative predominance of migrants from Georgia in the retail trade and of migrants from Armenia in health care is not sufficiently strong to allow any far-reaching conclusions.

Thus, these findings provide additional arguments supporting the hypothesis regarding the relative stabilization of Moscow's labor market. The major reason for this development relates to the shift from the predominance of fully unregulated migration to Moscow during the nineties to the efforts of the government on the Federal and local (Moscow) levels to regulate the process in an economically sustainable and more or less civilized way. These efforts are often subject to criticisms due to their alleged or real inefficiency. One should not forget, however, that the phenomenon of trans-boundary immigration is relatively new for Russian society. On the contrary, Russia has the historical experience as an emigration society. One may expect that the experience already gained in the steering of immigration will contribute to the better understanding and management of these processes.

The better understanding and management of immigration is essential in order to cope with the influential xenophobic attitudes in Russian society. Given this undeniable fact, primary information from the field studies reveals promising trends in the development of a mutual accommodation of migrants with the local population. The representatives of the first wave of mass labor migration to Moscow typically worked in firms owned largely by other immigrants, and more often than not with their compatriots and representatives of nationalities other than the local majority. The information obtained from the current labor migrants shows a substantial change in this respect. Whoever the owner of the firm, it is clear that the ethnic mixture of employees has already taken place. Under these conditions it is natural that most employees in the firms under scrutiny are citizens of the Russian Federation:

Figure 6: Most colleagues at your work place/business partners were/are:

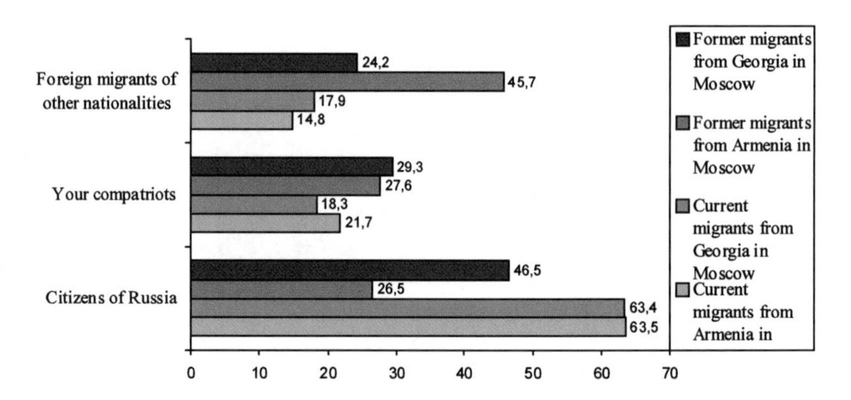

Overcoming substantial differences in the affiliation of migrants from Armenia and Georgia to firms with varying ethnic compositions of their staffs is still another striking outcome of the field study. It is indicative of this new situation of ethnic affiliations that the current migrants from Armenia and Georgia are working in firms with staffs comprised of strikingly similar ethnic compositions. Colleagues at the workplace are mostly Russian citizens, according to 63.5% of the current migrants from Armenia and 63.6% of migrants from Georgia. An explanation for this could expand into two different directions. On one side, firms in

Moscow have become increasingly open to the migrant labor force, which brings about a mixture of ethnicities in these firms with the natural predominance of the local ethnic majority. On the other hand, migrants from Armenia and Georgia in Moscow are generally moving away from ethnic enclaves and joining circles in which the local ethnic majority predominates. Whatever the moving forces of both processes, the effect is the more efficient inclusion of the migrants' labor force in the reproduction of Moscow's economy.

There are still other indications of the normalization of the status of migrants in the local labor market. The field studies registered an improving correspondence between the level and type of education attained and the character of migrants' occupational activity in Moscow. This is a universal problem in transboundary migration, usually affecting the employment of immigrants in labor market segments, which require a level of education below their own educational achievements. This effect is widely discussed as the phenomenon of "brain waste". And such was largely the case with the employment of migrants from Armenia and Georgia during the first large wave of migration to Moscow after 1990. Migrants from Georgia were more often affected by this discrepancy between the type and level of their education and the work they had to perform. But the information obtained from the currently employed and self-employed migrants from both countries in Moscow shows a somewhat different picture. Being better informed about employment opportunities in a labor market which is getting increasingly transparent, the migrants from Armenia and Georgia who were recently interviewed in Moscow have achieved a very similar and fairly high level of correspondence between the type and level of their education and the requirements of their occupational activity:

Figure 7: Does (did) your employment/self-employment/business in Moscow correspond to your education?

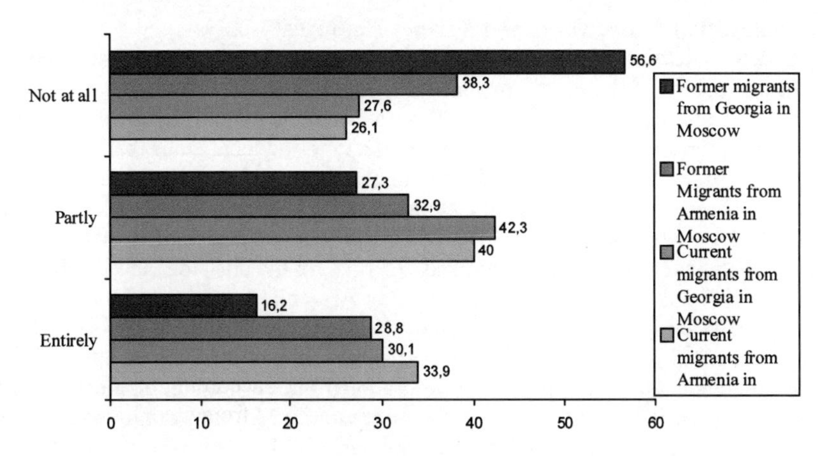

The changes in the conditions of employment are closely related to changes of the macroeconomic situation in Armenia and Georgia, as well as in the Russian Federation. Conterminously, the intentions and the status of labor migrants from Armenia and Georgia in Moscow have also changed. During the nineties labor migration to Moscow was regarded largely as a means of securing the survival of the migrants themselves and their families. Families that remained in Armenia and Georgia greatly relied on remittances that had to be sent regularly. The migrants from Armenia and Georgia who were interviewed in Moscow in 2009 had already modified their relationships with their countries of origin and their relatives in them. Most interviewed migrants had been settled in Moscow for a relatively long period of time. Half of the respondents from both Armenia and Georgia no longer felt an essential obligation to support relatives in their countries of origin:

Figure 8: Do you (did you) send financial support to your family, relatives or friends in your country of origin regularly?

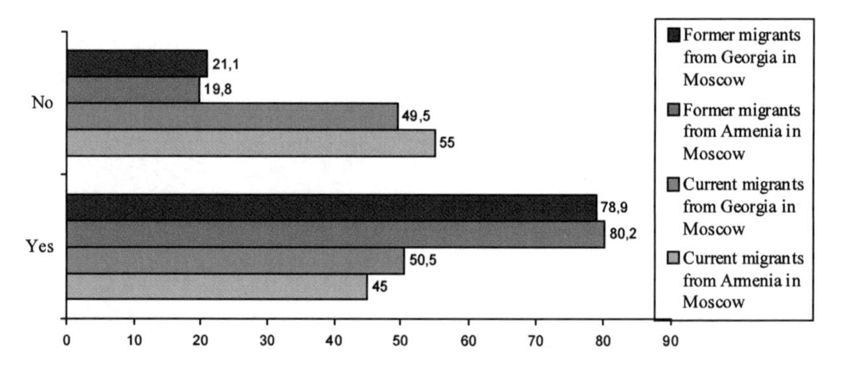

5 Conceptual Frameworks and Further Empirical Analyses

The brief analysis and interpretation of empirical findings above was focused on just one aspect of the multidimensional topics related to trans-boundary migration. The discussion covered the achievements and problems of the adjustment of labor migrations from Armenia and Georgia to the local labor market and labor conditions in Moscow. The following chapters will continue this analysis by taking many other relevant details into account.

The **macro-social** parameters of the process will be specified with a view to the local conditions in Armenia and Georgia, on the one side, and in Moscow on the other. The authors of the chapters on Armenia and Georgia largely make use of data obtained from the interviews with migrants and experts, as well as from the monitoring of events in order to establish the key characteristics of the *macro-economic* conditions for out-migration from each of the countries. This has been done not only to focus on the migration to Moscow, but to many other destinations, as well. The chapter covering data on Moscow goes into the details of the economic situations of various types of migrants from Armenia and Georgia. *Macro-political* conditions are the subject of a particularly detailed analysis from the point of view of the legal regulations of out-migration from Armenia and Georgia and the legal regulations of immigration from both countries to the Russian Federation. One may notice that all these regulations were legally non-existent or practically not applied in the 1990s and have been changed many times since their introduction. The discussion of the *macro-cultural* parameters of trans-boundary migration establishes a broad range of both cultural commonalities and increasing cultural differences between post-Soviet countries. This development is generated through the independent states' stabilization processes, each containing a plethora of economic and political specifics.

It is not the intention of the present study to specifically focus on the **meso-social** parameters of trans-boundary migration. A detailed study of this broad field would require special research instruments and a particular organization of research. Nevertheless, the *meso-economic* conditions of migration have already been touched upon in the context of the underdevelopment of agencies supporting migrants' employment search in Moscow. Some *meso-political* issues will be dealt with later on in the context of the legalization of migrants' residency and their efforts to obtain work permission. *Meso-cultural* conditions are included in any discussion of the activities of ethnic and religious organizations for migrants from Armenia and Georgia in Moscow.

The major focus of the conceptual development, operationalization, collection and processing of empirical information and its theoretically guided interpretation in the ArGeMi project relates to the **micro-parameters** of cross-boundary migration in the post-Soviet space. The *micro-economic* conditions of migration have been studied in parallel in both Armenia and Georgia, as well as in Moscow. The type of employment before and after migration, level of income, composition of households and the size of remittances sent back to the country of origin were dealt with in various contexts. The *micro-political* conditions were interpreted and studied in connection with the responsibilities of migrants to their families. Another issue of special interest concerned their assessment of the implications of cross-boundary migration for their country of origin. A crucial political decision of migrants in the host country is the decision whether or not to apply for citizenship of the Russian Federation. The empirical findings are rather indicative in this respect, documenting the widespread readiness of migrants to become citizens of the Federation. *Micro-cultural* conditions were empirically studied and analyzed in the context of the level of education of migrants, command of languages, their interests in specific information channels, etc.

As seen in a broader perspective, the experience of migrants in Moscow underlines the conclusion that the global economy cannot function efficiently without the commoditization of labor or the trans-boundary mobility of labor. The outlined problems of migrant labor are not unique to the Russian Federation. Global labor markets should most likely carry out a process of civilizational upgrading. In the European Union most conditions for the legal and institutional integration of migrant labor already seem to be quite well developed. Nonetheless, migrants' work in the shadow economy and difficulties with their integration into host societies is permanently on the agenda of political discussions and legal decisions in the Union. However, the institutional framework of the Russian Federation is still far from this pattern of a well-regulated social market economy. The visible improvement of the legal status of migrants in Moscow notwithstanding, one can still hope that a growing understanding of the contribution of the migrants to the wellbeing of Russian society will strengthen the economic, legal and moral respect for this very important segment of the labor force in the Russian Federation, and Moscow in particular.

References

BUCHANAN, Jane (2009): *„Are You Happy to Cheat Us?" Exploitation of Migrant Construction Workers in Russia.* New York: Human Rights Watch

EUROPE IN FIGURES – *Eurostat Yearbook 2008* (2008) Brussels: European Commission

GALKINA, Tamara A. (2006): 'Contemporary Migration and Traditional Diasporas in Russia: The Case of Armenians in Moscow'. *Migracijske i etničke teme,* Vol. 22, N 1-2, pp. 181-193

GRIGOR'EV, Maxim and Andrei OSINNIKOV (2009): *Nelegal'nye migranty v Moskve* [Illegal Migrants in Moscow] Moskva: Evropa

IVAKHNYUK, Irina (2009): *The Russian Migration Policy and Its Impact on Human Development: The Historical Perspective.* UNDP

LEVADA-CENTR: Press release of 07.12.2009, http://www.levade.ru/press/2009120704.html (in Russian, 15.08.2010)

MARX, Karl (1970) [1847]): *Wage Labor and Capital.* Moscow: Progress Publishers

Migrants "distort economy" – Moscow's mayor. RIANOVOSTI, 18.09.2010, http://en.rian.ru/Russia/20100918/160635154.html (30.09.2010)

SASSEN, Saskia (2007): *Sociology of Globalization.* New York: W.W. Norton & Company

SASSEN, Saskia (2010): 'Global Inter-city Networks and Commodity Chains: Any Intersection?' *Global Networks,* Vol. 10, N 1, January, pp. 150-163

Sovremennaya demograficheskaya situaciya v Rossiyskoy Federacii [Current Demographic Situation in the Russian Federation] (2010) Moscow: Federal Office of State Statistics http://www.gks.ru/free_doc/2010/demo/dem-sit-09.doc (29.09.2010)

STALIN, Josef V. (1952): *Economic Problems of Socialism in the U.S.S.R.* Moscow

The Urban Elite: The A.T.Kerney Global Cities Index 2010 (2010): Chicago: ATKERNEY http://www.gzt.ru/upload/files/10/10/05/1286292536785.pdf (07.10.2010)

VISHNEVSKIY, Anatoliy, et al. (2010): 'Formirovanie migrantskih segmentov v ėkonomike' [The Formation of Migrant Segments in the Economy]. *Demoskop Weekly*, N 431-432, 23.08.-05.09.2010 http://demoscope.ru/weekly/2010/0431/tema03.php

World Development Report 2010: Development and Climate Change (2010). Washington, D.C.: The World Bank, pp. 378-379

ZAYONČKOVSKAYA, Žanna (Ed.) (2009): *Immigranty v Moskve* [Immigrants in Moscow]. Moskva: Tri Kvadrata

ZAYONČKOVSKAYA, Žanna, Nikita MKRTČYAN and Elena TYURYUKANOVA (2009): 'Rossiya pered vyzovami immigracii' [Russia Facing the Challenges of Immigration]. In: Zayonchkovskaya, Zhanna and V.S.Vitkovskaya (Eds.): *Postsovetskie transformacii: otrazhenie v migraciyakh* [Post-Soviet Transformations: Effects in Migrations]. Moskva: IT "Adamant", pp. 9-62

Gevork Poghosyan

Out-Migration from Armenia

1 Introduction

The 20th century was rightfully called "the epoch of migrations" (Castles and Miller 1996: 43). The amount of international migrants has increased from 75 million in 1965 to 120 million in 1990 and reached the level of 140 million at the turn of the last century. The amount of migrants in the world increased faster than the growth of the earth's population (Yudina 2002: 102). As a consequence of the migration increase, modern societies became culturally, ethnically and religiously more diverse. The main flow of migrants (42.7%) is directed towards Europe, Northern America, Australia and New Zealand, with Belgium, France, Germany, Italy, the Netherlands and Switzerland as main destinations in Europe.

After the dissolution of the Soviet Union and the emergence of 15 independent states in 1991, the marketization of the economy and the globalization caused massive out-migration from the former Soviet republics in Southern Caucasus. The out-migration was fostered by ethnic conflicts as well as by social and economic problems there. During the last decade of the 20th century, Armenia, Azerbaijan and Georgia lost about three million inhabitants in out-migration. In addition to the increased international migration the internal migration also increased significantly. According to the data of the European regional bureau of UNHCR, in the space of the *Commonwealth of Independent States* (CIS) already in 1991-1992 about 700,000 refugees and 2.3 million of internally displaced persons (IDPs) were registered (Castles 1999: 7).

In 1999, the *International Organization of Migration* (IOM) issued its special report "The Migration of the population in the CIS countries: 1997-1998". The authors stated: "The main migration trend in 1997 on the post Soviet territory was the movement to/from Russia. In 1997, comparing with the previous year, the immigration to Russia from CIS countries and Baltic countries decreased only for 7.5%, whereas the emigration from Russia decreased for 20%. Russia is the main point for the majority of emigrants from the CIS countries. While the superiority of emigration over immigration is still observed in the majority of CIS countries, for Russia opposite tendencies are typical" (Zayonchkovskaya 1999: 14). The majority of all intra-CIS migrants are Russians (54%), followed by Ukrainians (16%), Byelorussians (3.5%) and migrants from the South Caucasus (6%) (Zayonchkovskaya 1999: 14-15).

Compared with the worldwide migration trends, there are certain specifics in migration processes from the South Caucasian region. The main results of massive emigration during the last two decades from Armenia are depopulation and mono-ethnicization. At the turn of the 20th century, about one million people had emigrated from Armenia. The main reasons which caused such an exodus were the abrupt reduction of work places, the economic crisis and the immense deterioration of living conditions. Armenia became an exporter of workforce and particularly of skilled workers. The emigration of the economically active population on a large scale essentially changed the demographic structure of Armenia's society.

2 The Local Experience of Out-Migration

There were several massive migration flows in Armenian history. In the 20th century Armenians, together with other ethnic groups in the country migrated because of genocide, deportation, political and religious pressure, wars, bad economic conditions and crisis, social conflicts and natural disasters. Except of the 1915/16 genocide, the reasons for migration in the 20th century did not differ from those in previous centuries (Karakashian and Poghosyan 2003: 227).

According to data from the *National Library of Armenia*, in 1917 there existed 202 migration settlements ("colonies") of Armenians spreading from Madras in India to Australia, Argentina, Ethiopia, North America, to European and Russian cities. There is also information about the emigration of Armenians from the Ottoman Empire to the USA and Europe that took place until 1895. Until 1915, the majority of Armenian population was rural or provincial, with numerous specialized craftsmen. But there were also successful Armenian traders, merchants and entrepreneurs in the provincial centers and in particular in the Ottoman capital city of Constantinople (Istanbul).

Repeated massacres in 1894-96 and 1909 and the genocide of 1915/16 caused massive flight of the survivors. It was estimated that during 1894-96 200,000 Armenians died from massacres and starvation in their aftermath (Karakashian and Poghosyan 2003: 229). About 100,000 Armenians found sanctuary in other countries (Dadrian 1995: 112), not only in the neighborhood, but also in Trans-Atlantic destinations. In 1895-1899 about 70,980 Armenians left for USA (Tashjiyan 1947: 18).

The genocide and forced deportation of Armenians from West Armenia and other areas in the Ottoman Empire during the First World War, which was committed by order of the ruling *Ittihat ve Terrakki Cemiyeti ("Committee for Union and Progress")*, or so-called Young Turkish government, was followed by the dispersion of the survivors to different countries. To a large extent, the present-day Armenian Diaspora emerged as a result of that genocide.

Armenian Diasporic settlements existed since the loss of Armenian statehood in the 11th century. But they grew significantly in size and number after the Genocide. Some 1.5 million Armenians perished, others managed to escape and established communities in Eastern Europe, the Balkans, and in the Middle East.

Thousands of Armenians settled in Western Europe and in both North and South America. Substantial Armenian communities emerged in the Russian Far East as well as in the former Soviet republics. Armenian communities can also be found in India, Australia, New Zealand, Sub-Saharan Africa (Sudan, South Africa, and Ethiopia), in Singapore, Myanmar and Hong Kong. Armenian exile communities even once thrived in China, Japan and the Philippines. According to experts, the amount of Armenians in the world is estimated between 8 and 9 million (Hayadaran 2009: 594; Migration in Armenia 2008: 23; Panossian 2003: 12; Migration Trends in Eastern Europe 2002: 51; Sanoyan and Epstein 2010: 66). Today, the countries with the largest number of Armenians (excluding Armenia and Nagorno-Karabakh) in rank of order are Russia, the United States, Georgia, France, Ukraine, Iran, Lebanon, Syria, Argentina and Canada.

The true amount of the Armenian Diaspora in many countries of the world is still under-researched. Statistical 'hard figures' are difficult to established due to ongoing migration processes inside the Armenian Diaspora. For example, as a result of conflicts in the Middle East and the low human rights standards there, thousands of Armenians emigrated from Turkey and the Middle East countries in order to establish themselves in the United States, Canada, France and elsewhere.

A massive wave of 'repatriation' to Armenia started in the Middle East in 1946. At that time the government of Soviet Armenia invited Armenians of the Diaspora to come to live in Armenia. Many people of different professions in Lebanon, Syria, Palestine, Egypt, Greece, Iran and France were selling up their property and made their ways towards their historical homeland. The total amount of foreign ethnic Armenian immigrants to Soviet Armenia in the 1921-1962 period was estimated at 220,000, and for the years 1962-1973 at 26,140 (Koutcharian 1989: 196).[1] The 'repatriation' climaxed during the immediate post-war years of 1945-48, when about 90,000 Armenians immigrated to Armenia from twelve countries (Hay spyowrk' hanragitaran 2003: 15)[2], but declined sharply after 1965.

Due to Armenia's geostrategic position and the circumstances caused by this position, Armenians had always to be mobile and to develop qualities that allow them to adapt to new habitats. However, under Soviet rule their mobility was artificially restrained. The Soviet regime harshly limited the freedom of movement, especially out-migration. With the collapse of the Soviet regime, a 'migration spring' started.

---

1 Slightly divergent figures are mentioned by other authors. From 1920 to the 1980s, about 170,000 ethnic Armenians immigrated to Soviet Armenia. In 1920-1936 they were some 42,200 people, in 1946–1948 89,700 and in 1962-1982 32,500 (Hayadaran 2009: 593).

2 E. Khatanassian mentions a total of 178,000 immigrants in the 1945-65 period (Khatanassian 1965: 42).

The labor surplus is a relevant reason for the Armenian out-migration. Already in Soviet times, Armenia was a country with larger labor resources than could be actually provided with work. Seasonal migration of Armenians was widespread. Currently Armenians continue to make their way to Russia and to other countries of the former Soviet Union in search of earnings (Khojabekyan 2004: 209-212). The major destinations are places where they have already established business contacts and relationship with the local Armenian Diaspora. "Somehow exceptional is the high amount of arrivals from Armenia despite the fact that Armenia is no direct neighbor of Russia (19,123 arrivals in 1997 related to 30,751 in 2007). Among the CIS states, Armenia has the highest share of her workers abroad (perhaps 700,000 in a labor force of 1.2 million)" (Savvidis 2009a: 150).

Economists believe that in the future Armenia will produce an annual labor surplus of about 70,000 to 80,000. "Provided that the economic growth will be 6% per year and can create 30-35 thousand work places, then even under these circumstances the home market will not be able to absorb the incoming labor force within the next few years" (Papoyan and Bagdasaryan 1999: 9). The academician Khojabekyan notes that during the 1986-1990 period Armenia's workforce was just one million on average. This figure grew to 1,350,000 people already in 1996-2005. Thus, the number of persons who are in need of job placements increased for more than 25%.

The waves of forced emigration of Armenians from Azerbaijan began in 1988. Due to slaughters during 1988-1990 many found refuge in Armenia, Russia, USA and other countries. According to data of UNHCR, some 264,339 refugees from Azerbaijan moved to Armenia (Statistical Yearbook 2001: 27). During the first decade of the 21st century, about 100,000 of them became citizens of the Republic of Armenia. The remnant left Armenia in search of earnings and better living conditions elsewhere. The adaptation of the refugees lasted for several years. Armenia's society and state were not prepared to solve a problem of such dimensions. Nevertheless, the authorities eventually determined their policy concerning the refugees and headed on towards their full integration.

During the decade between 1990 and 2000 almost 900,000 people emigrated from Armenia (Migration Trends in Eastern Europe 2002: 51). This is one third of the current population of the country. Hundreds of thousand migrants from Armenia made their way to Russia, Ukraine, Belarus, USA, Europe and remoter destinations. In Central Europe, Armenian migrants tried to settle down in Germany (Poghosyan 1997: 20), Belgium, Netherlands and Poland (Marciniak 2001: 107-116). The immigration to Germany remained small and unsuccessful. Most asylum applications failed. In recent years, there is some migration to Austria and more so to France, Spain and Greece. In some cases the migration flows of Armenians are directed to the countries where relatively large groups of the historical Armenian Diaspora already exist. Many of Armenia's out-migrants have departed without any preparation of documents before departure. Initially, they assumed to leave not for a long time to earn some money and then to return. But the large majority continues to live in different countries as irregular or even illegal migrants. Many others belong to the increasing category of seasonal migrants.

The social-demographic structure of migrants includes people from all strata of Armenian society. The majority of them (65%) are men of working age. Their education level is high. As a rule they are qualified workers with average and professional-technical education. A significant share has a completed higher education (30-35%) (Human Development Report Armenia 1999: 46). In the 1991-1994 periods, male migrants were leaving as singles and tried to settle down abroad. Later on they started to leave with their families.

Armenia belongs to those countries that loose qualified specialists and intellectuals. It is destructive for a small country to spend means on educating specialists and then to 'export' them in a significant number. With this brain drain Armenia loses intellectual and spiritual wealth that has been developed for decades.

Another effect of out-migration is the contribution of migrants to the economical survival of their families and friends at home by remittances. At the same time, the migrants' absence harms the economy of Armenia. They deprive a potential Armenian class of businessmen of markets and slow down the formation of a middle class in the country.

Today new migration processes are taking place, in which the state still does not take any relevant part. One of these processes concerns the return of migrants to Armenia and their reintegration. Another process is the trafficking of women to the Arabic Emirates, Turkey and Greece for exploitation in the sex-industry (Poghosyan 2005b: 23). Finally, there are also transit migrants within Armenia (Poghosyan 2009b: 193-212). Refugees from Afghanistan, Pakistan, India, Iran, Turkey and even African countries have arrived in Armenia. They are part of the global flow of migrants from the South to the North. Armenia is a buffer state, from where the transit migrants try to leave for Western Europe. Kurdish refugees have crossed the Armenian-Turkish border and were arrested in the international airport when they were trying to depart to Amsterdam (Karakashyan and Poghosyan 2003: 239). In 1998-2000 some 232 transit migrants were registered in Armenia (Migration Trends in Eastern Europe 2002: 51-52). The majority of them fails to move further and returns to their countries of origin with the help of international organizations.

The analysis of migration can provide a detailed classification of the motives and circumstances that caused the outflow of the population which occurred in Armenia during the last 18 years (Kharatyan 2003: 9). The reasons differ widely. As Nikolai Genov notes, "International migration is a phenomenon categorically rejecting any kind of mono-causal explanations. Material structures and value-normative orientations are simultaneously involved in decisions to emigrate and in the efforts to accommodate in the country of immigration. Individual preferences and calculations together with collective patterns of orientation and action mix in the decision to take the path of emigration. Which of these factors might be more influential in determining international migration might be matter of specific empirical proof but not a matter of a generalized analytical decision" (Genov 2009: 171).

Researchers identified the main reasons for out-migration to be the war and blockade of Armenia by Azerbaijan since 1989 and by Turkey since 1993 together with the subsequent energy crisis, cold and starvation in the early 1990's (Poghosyan 2005: 206-208). But there are other 'driving' factors, too.

Cultural-historical factors: Here we may refer to the numerous Armenian Diasporas in many countries of Europe, Asia and America. The presence of the Armenian Diaspora is surely an important factor that stimulates the migratory disposition of the people, taking into consideration the closeness of the contacts and relations that are typical for Armenians.

Political-historical factors: First of all we shall refer to the collapse of the Soviet Union and the emergence of 15 new independent states observing the principles of the freedom of movement. The process is called "anti-camp" syndrome related to the fact that the citizens of the previous socialist camp member states suddenly obtained the right to travel freely. Longstanding Western political propaganda of appraisal of Western values, liberties and the Western way of life was addressed to the Soviet population. It should also be counted among those factors. Many Soviet people were influenced by such propaganda and made their ways to Europe in search of better living conditions.

A special domestic political situation emerged in the new independent former Soviet republics when new national political elites came to power. In the case of the three South Caucasian states the focus on national issues such as language, culture, traditions etc. was typical. The Russian-speaking community felt largely ousted and opted for emigration due to nationalist or even chauvinistic processes. The ousting of the Russophone community partly explains the tremendous difficulties in the integration of ethnic Armenian refugees coming from Azerbaijan to Armenia. This particularly concerned the urban populations from Baku and Sumgait. Many of these immigrating Armenians were Russophones and also highly qualified specialists. The national 'renaissance' in Armenia during the early 1990is created discomforts for these refugees and triggered their migratory disposition.

Still another important political-historical factor of migration was the change of the political system. This factor is mentioned in the literature least of all. But it is quite relevant. During the Soviet times, large groups of qualified cadres emerged in Armenia and elsewhere. For the majority of them the collapse of the Communist regime meant the end of their work career and a profound change of their life strategies. These people found themselves in the position of social outcasts in the new society. Many of them emigrated to Russia and to other countries at the first opportunity they had.

War and conflict factors: These factors caused multiple crowds of refugees and internally displaced people (IDPs) in the country. The armed conflict over Nagorno-Karabakh gradually developed into an undeclared war of Azerbaijan against the Armenians of Nagorno Karabakh. The war caused the displacement of a huge amount of people on both sides of the conflicting parties. About 500,000 Armenians were forced to escape from Azerbaijan. This number included some 48,000 temporary refugees from Nagorno-Karabakh who could return there after some years. The majority of the Armenian refugees, approximately 360,000, were directed to Armenia. Some 72,000 inhabitants of Armenian villages situated near the border to Azerbaijan were forced to leave their homelands temporarily because of constant shelling and insecurity (Statistical Yearbook 2001: 27; Khojabekyan 2004: 233; Migration Trends in Eastern Europe 2002: 51). In 2006 the number of refugees in Armenia was 113,714 (Migration in Armenia 2008: 15). The depressing situation of war and long lasting conflict became the main reason for the emigration of thousands of people from the country

Economic factors: Armenia found itself confronted with a wide range of harsh economic conditions like blockade of international roads and communications, suspension of almost half of the industrial enterprises of the country, a heavy energy crisis, high level of unemployment, abrupt drop of the living standards for the majority of the population, loss of former markets and of providers of raw materials. The break-down of Armenia's economy became the main migration reason of almost one third of the population in recent years and in particular in the last decade of the 20th century. The Armenian official statistic indicates increasing unemployment during the period 2007-2009 due to the impact of the global financial and economic crisis. The number of officially registered unemployed receiving compensation from the government in January 2009 was 16 700, while in January 2010 their number had increased to 26 300, which makes a 58% increment (Sotsial'noe polozhenie 2010: 36). As compared with 2009 and according to the same official statistics, in 2010 the quantity of crimes in the country has increased by 42.1% (Sotsial'noe polozhenie 2010: 71). Unemployment is one of the reasons of the crime growth.

Psychological factors: At different stages the impact of these factors weakened or intensified. But as a rule the crucial issue became the loss of social self-esteem which influenced migratory dispositions. For a certain part of the population, the psychological factors are first of all connected with the awareness of some loss of 'tranquil life' and the desperate present situation. People did not see a better future for their children and tried to establish their life abroad. The psychological factors of despair and uncertain future are important push-factor of out-migration (Poghosyan 2003a: 116-124; Arakelyan, Haroutyunyan and Poghosyan, 2007: 133-155; Poğosyan 2006a: 236).

These are the most relevant but by far not the only reasons for mass post-Soviet out-migration from Armenia. The out-migration of Armenians is a multidimensional phenomenon. Sometimes it is wrongly simplified and limited to only few obvious reasons.

3 Aims and Methodology of the Field Study

The ArGeMi project covers the whole cycle before, during and after migration. In correspondence with the project's objectives, the Armenia team conducted surveys before, between and after travels abroad in order to explore motifs, forms, courses, results, potentials and perceptions of the observed cases of migration. Special attention was paid to labor migration, to the structures of qualified labor migration, to practices of money transfer and remittances, as well as to the return potential. The diversity of survey methods was applied for exploring the experience and expectations of migrants, migration related experts and main stakeholders of the post-Soviet migration process. Following methods were applied in the ArGeMi project in Armenia:

- Surveys before, during and after the migration cycle in Yerevan (capital and largest city), Gyumri (second largest city), Vanadzor (third), Goris in the South of the country and Alaverdi in the North;
- Monitoring of events related to out-migration;
- In-depth interviews with experts and other related stakeholders.

The field study was conducted during February, March and April 2009, when seasonal migrants from Armenia normally leave the country. The survey was conducted among a sample[3] of 300 returnees from Europe, Northern America and Asia, 200 returnees from Moscow and 100 potential migrants. The sampling technique applied was the snowball procedure of the identification of respondents. The interviews with 20 experts were conducted among political administrators and decision-makers, representatives of international bodies, human rights non-governmental organizations (NGOs), scientific experts and organizations of migrants. The interviews revealed different degrees of problem awareness, flaws, failures or even a complete lack of migration regime and legislation, and a general weakness in statistics. The interviews of the experts were carried out within the three months of February until April 2009. All experts were selected according to their involvement into Armenian migration problems for several years.

The monitoring of out-migration related events was conducted with special attention to out-migration from Armenia to Russia and particularly to Moscow. 'Event' is defined as an occurrence, significant happening, or outcome of a previous phenomenon. Events were identified and analyzed according to information from official documents of the state or other actors, mass media reporting and scientific studies. The monitoring of events was carried out twice: The

3 Called henceforward "other destinations"

first round of migration related event monitoring was conducted in Armenia during the February-April 2009 period, while the second round took place between January and March 2010. Both monitoring rounds covered three months each. A particularly challenging issue related to the monitoring was the distinction between the manifest appearance of the event and its hidden causes and latent meaning. Three main Armenian newspapers – the official *Hayastani Hanrapetutyun* (in Armenian), the independent *Golos Armenii* (in Russian) and *Aravot* (in Armenian) as well as two main Armenian public TV stations 'H1' and 'H2' were monitored on migration events. In total, twenty newspaper articles and eight TV programs were analyzed during the first round of monitoring of events and 25 newspaper's articles and eight TV programs during the second round.

4 Recent Trends of Out-Migration from Armenia according to Experts

The common pattern in the answers of interviewed experts was that none of them possessed exact information concerning the amount of migrants, particularly of those who departed to Russia in general and to Moscow in particular. The dominant opinion of experts was expressed by the representative of the IOM office in Armenia as follows: "According to the official data and the results of independent studies, 800 thousand to one million people have left Armenia since 1991. There is no exact statistics in relation to how many of them reside in Russia and particularly in Moscow. However, taking into account the fact that the main flow of the migrants was to Russia, we can suppose that 60-70% reside there". Some of the experts (members of local NGOs) mentioned that approximately 500 thousand Armenians live in Russia, while about 150 thousand to 200 thousand of those live in Moscow.

All experts mentioned the amount of roughly one million émigrés from Armenia. Meanwhile, in the opinion of the experts' majority the overwhelming part of Armenian migrants leave for Russia and there, as a rule, to Moscow. This closely corresponds with the results of a sociological survey conducted in 2007: "The Russian Federation continues to be the most popular country of destination for Armenian labor migrants (93%)." The survey also reported that most of the migrants (39.8%) looked for jobs in Moscow (Minasyan et al. 2007: 25-26).

In the ArGeMi survey experts named a wide range of estimates, although agreeing that the vast majority of migrants from Armenia go to Moscow. Thus, in the opinion of a sociologist from the Yerevan State University, probable one million Armenians reside in Moscow, the majority of them unregistered. In other expert's opinion it is difficult to ascertain the exact number of migrants in Moscow because the data of the official statistics of the Russian Federation cover only a small part of all migrants from CIS countries, Armenia included. The majority of migrants are assumed to live there without official registration. But all experts agreed on the fact that the majority of migrants from Armenia in Russia live in Moscow. Some experts with scientific backgrounds referred to data published in Armenian mass media which mentioned that in 2008 about 50,000 migrants went from Armenia to Moscow for labor.

The experts were also unanimous in the opinion that 60-70 % of all migrants from Armenia were males of working age: "I think women account for nearly 40 percent or less, well, certainly the rest are men, about 60 percent and more" /Academician/. This opinion coincides with the published data of surveys conducted in Armenia at different times (Zayonchkovskaya 1999: 14-15; Migration in Armenia 2008: 23; Human Development Report Armenia 1999: 46; Poghosyan 2003a: 195; Poghosyan 2005b: 59). These data also coincides with those of the ArGeMi project survey, according to which 67.5 percents of the interviewed returnees from Moscow were male.

In the experts' opinion the majority of migrants have secondary, secondary vocational and incomplete or complete higher education. Thus, migrants from Armenia have a quite high educational level. A representative of the OSCE office in Armenia mentioned: "Previously mainly smart, educated people and scientists left. These were the ones who left in 1994. Today mostly seasonal laborers leave, i.e. common workmen. There are lots of seasonal laborers." This opinion is confirmed by the data of several surveys including the ArGeMi survey with returnees from Moscow. According to that data, 29 percents of the respondents had higher education, 40 percents secondary special or vocational education. Some experts believe that the migrants from Armenia in Moscow work mainly on the construction sites. "Migrants from Armenia are mostly occupied in the sectors of construction and service. They are occupied in the construction sector not as cheap workers but as masters. They are usually working for Armenian contractors. The migrants who are occupied in the service sector are not seasonal migrants, but have asserted there for some time, for example in car service or restaurants" (Sociologist from the Yerevan State University).

According to the interviewed experts, the local population in Moscow "has a negative attitude towards the migrants from Caucasus. Of course, their rights are violated. There is murder, violence because of national identity. Migrants are unprotected" (Academician). Some experts even mentioned that there are numerous incidents of open aggression towards migrants: "I think that there is no special treatment of those migrants from Armenia who came to Moscow after the 1990s. Skinheads do not look into passports. The treatment is one and the same for all ethnic groups. That is a nationalistic approach. They may confuse Tajik people with Udmurts and kill them. The Russian President says that they do not act against Armenians and he also says that Armenians are our relatives. Everything is OK on state authority level, but in practice no attention is paid" (Journalist). The experts' agreed that the labor rights of migrants in Moscow are poorly protected. "The rights of those who illegally stay in Moscow are not protected. Their passports are taken and they become victims of trafficking. The rights of those who live there illegally are protected only as much as the rights of the local residents are protected" (President of the Women Union of Armenia).

The experts were not optimistic about the return of many migrants to Armenia. They rather believed that there is no obvious return flow back and that the situation in Armenia has not improved sufficiently enough for economic migrants to return. In the interviews the returnees from Moscow and from other destinations expressed the same skepticism.

Almost all experts could provide information about remittances from the migrants working in Russia. Although they mentioned different amounts, the estimates were quite impressive sums. "The remittances are estimated between 1.7 and 2 billion US$. This amount was registered in 2006-07 and it was a record" (Academician). According to another opinion, "before 2009 the amount of migrants' remittances was annually 2 billion US dollars. Due to the crisis, there will be serious changes here and the data need to be updated. Russia has seriously suffered from the crisis and it is natural that the migrants in Russia earn less than before. Consequently, the amount of remittances declines. Therefore, today it is necessary to talk about a 34% decrease in remittances. Most of the money comes from Moscow since it is the main destination of out- migration from Armenia" (Representative of the IOM mission in Armenia).

These assessments correspond with the opinion of experts in Russia. "The estimation shows that the situation in Armenia is difficult … Today already 70% of the population feels that the crisis affected the economy of the country… The situation in the republic was aggravated by the fact that an essential stake of the GDP of the country was formed by the means transferred by the representatives of the Diaspora. The worsening of the financial standing of the representatives of Diaspora seriously and negatively affected the Armenian economy. The other factor that affected the economic situation in the republic was the problem of return of the Armenian workers who worked abroad. Armenian sources do not bring the exact statistics on the number of those who returned but it is supposed that there are a considerable number of them. Those categories of Armenian citizens join the dole queue and increase the social expenses of the government" (Grinyaev 2009: 10-23).

According to the head of the *Migration Agency* at the Government of The Republic of Armenia, for the first time a positive balance of persons leaving and returning to Armenia was observed in March 2009. This is the month when seasonal migrant workers typically return to their jobs abroad after a 'hibernation' period at home. The number of Armenians leaving the country for work decreased by 56% in spring 2009 compared to their number in the previous year. The reported reversal of the trend in Armenia's outward migration during the typical departure period of labor migrants naturally means that more people in Armenia were in search of jobs in 2009 after they lost job opportunities abroad. The monitoring of events in January-March 2010 verified the expectation of the experts. The rate of under-employment according to the officially registered cases increased from 75,200 in 2009 to 84,700 in the first decade of 2010. During the same period thousands of labor migrants did not leave for work in Russia because numerous workplaces were cut there due to the crisis.

"Migration is a very sensitive phenomenon", concludes the head of the Migration Agency. "Changing social, economic and political processes are reflected in migration phenomena as well. Naturally, a cataclysm like the current economic crisis could not but have an impact on it." Seasonal migrant workers who return in late autumn and leave for work abroad in early spring were the first to feel the impact. According to the expert estimated 95 percent of the total number of migrants from Armenia seeks jobs in Russia, with about 60-70 percent in construction. Meanwhile, Russia is affected by the deepening global crisis resulting in dwindling of construction and other sectors that used to provide jobs for foreign workers.

The estimates of Armenia's *Migration Agency* about the movement of people are based on information from the state border customs. As a rule, the largest negative balance of the cross-boundary movement is observed in the period from February to April. The negative balance of departures and arrivals in February-March 2009 comprised an increase by 46 percent as compared to the same period in 2008. Thus, a positive migration balance of 1,200 was reported for March 2009, with nearly 70,000 people who stayed and potentially joined the army of unemployed in Armenia.

Dwindling private remittances transferred to households in Armenia by family members or relatives working abroad appear to be another major consequence of the reduced "migrant worker export" in 2009, according to the head of the *Migration Agency*. The official statistics posted by Armenia for January-May 2009 showed a fall of ca. 40 percent in remittances against the same period in 2008. In the past five to six years, remittances grew by an average of 25 percent annually. The last quarter of 2008 showed a 20 percent (or more) decrease in remittances as compared to the preceding three-month period. But as a matter of fact remittances transferred to Armenia tend to increase at the end of the year. Already in the last quarter of 2008, remittances to individuals through bank transactions decreased from $530 million to $430 million. The decline of remittances continued throughout 2009. In January 2009, remittances decreased by 25 percent and in April by 39 percent. And again, 80 percent of the private remittances to Armenia come from Russia. In contrast to previous years characterized by rapid growth in passenger flows, this indicator has also fallen in 2009 according to the *Migration Agency*: "This year the flow of passengers has declined by 1-2 percent (or 7,600 people). Perhaps the number is not large, but the phenomenon shows that fewer people can afford to travel. This is also the consequence of the crisis".

5 Monitoring of Events

The monitoring of events was carried out twice from February to April 2009 and from January to March 2010 on daily basis. In 2009 the majority of the analyzed events were monitored on TV channels. The events mostly concerned messages about what had happened to migrants from Armenia in Russia. Other monitored programs covered interviews with experts on issues of refugees and the return of migrants to Armenia, or were special programs dedicated to trafficking and its Armenian victims. A separate program was about the Government of France that had assigned huge means for assistance to returnees.

Since February 2009, most newspaper articles covered the effects of the global economic crisis on migration from Armenia and the reactions of migrants. There was official newspaper information that remittances from Armenian out-migrants (mainly from Russia) decreased by 40% because of the crisis. It was also reported that for 2009 the Russian Government reduced the labor quotas by 30% for all migrants from CIS countries. As a result, many migrants from Armenia remained unemployed in Russia and were expected to return. Newspaper articles reported experts' estimates that the number of potential returnees to Armenia might reach 100-120 thousand people as a result of the crisis. For a republic that faces great difficulties in creating about 30 thousand new work places every year this would be too large number to accommodate in the national labor market. Subsequently, the number of unemployed would increase.

In newspaper articles experts mentioned that the economic crisis in Russia had reached its peak in those winter months when the migrants from Armenia return home as a rule in order to spend the holidays with their families. Therefore, a significant part of migrants from Armenia were caught by the crisis at home. The experts expected that they will not leave for Russia in the spring as the migrants did in the previous years. As a result the influence of the global crisis struck Armenian economy by both the reduction of remittances and by the increased unemployment at home.

In many newspaper publications in February-March 2009 three homicide attacks on persons of Armenian nationality in Moscow were debated. The murder of two Armenians on 7th February 2009 in Moscow drew a wide attention. A mother and her son were killed by unidentified bandits (Savvidis 2009b: 49). Almost all the newspapers in Armenia wrote with great indignation about this incident as a threat to the Armenian community in Moscow. The reaction of the state officials in Yerevan and the Armenian Embassy in Moscow was passive. This indifference was criticized in Armenian newspapers. The newspapers monitored during February-March 2009 addressed also several other criminal incidents the victims of which were persons of Armenian nationality in Russia in general and in Moscow in particular. After mid-March 2009 there was not a single publication or news on migration in the Armenian newspapers. The issues of migration were overshadowed by problems of the Armenian–Turkish relations and by events related to the 94th commemoration of the 1915 Genocide victims in Ottoman Turkey, which annually takes place on April 24th.

In 2010, the overwhelming majority of monitored events in the Armenian media were messages about the connection of migration and the economic crisis, Armenian migrants in Turkey, trafficking and return of labor migrants from Russia. During these three months the mass media continued to pay special attention to the situation of Armenian migrants in Russia and to the reduction of work places in Russia. Data of the *Central Bank of the Republic of Armenia* were reported concerning the fact that in 2009 the money transfers to Armenia decreased by 17% in comparison to 2008. The government of Armenia was criticized because in its anti-crisis program of 2009 it did not take into consideration the potential return of Armenian migrants. The issue was burning since the annual quotas of the government of the Russian Federation were reduced from six million in the previous years to three million in 2009 (*Aravot*, 24 March 2010). Meanwhile migrants from Armenia have made a major contribution to the economy of Russia. During the years preceding the crisis Armenians created about one million work places in Russia. In order to transfer one dollar from Russia to Armenia the migrants make deposit to the Russian economy equal to 7 dollars. The economies of the migrant receiving countries benefit from migration.

According to data of a sociological survey conducted by the Armenian representative in the Czech organization *People in Need*, 63% of 155 interviewed potential migrants from Armenia would like to leave for Russia (*Golos Armenii*, 18th February 2010). According to data of the ArGeMi survey conducted among 100 potential migrants from Armenia, 46% would like to leave for Russia, 14% to USA, 9% to France, while 28% mentioned other destinations. According to the data of both surveys the majority of all potential migrants intend to leave to Russia. The survey of the Czech organization also indicates that 60% of the migrants leave in search of a well-paid job and therefore should be defined as labor migrants.

During the second round of the event monitoring, the expectation was articulated that the decrease of work places and quotas for migrants in Russia caused a re-orientation of emigration to other countries. This assumption was based on the information of the UNHCR about a 43% (= 6,226 more applicants) increase of the number of asylum seekers from Armenia in different European countries in 2009. Most applications went to France, Belgium, Austria, the Netherlands, and Germany (*Aravot*, 27 March 2010). Again deadly accidents with Armenians in Russia were discussed. During the months of monitoring three TV reports on cases of murder of Armenians in Moscow and other cities of Russia were published. Also the problem of xenophobia in Russia was discussed at length in the media.

Another much discussed media topic was the notorious statement of Turkey's Prime Minister Recep Tayipp Erdoğan who in a BBC interview of 17th March, 2010 threatened to deport Armenian migrants from Turkey. The background of the statement was the resolutions of the Swedish legislators and the Foreign Committee of the US House of Representatives on the Ottoman genocide against Armenians, Pontos Greeks and Assyrians. In his interview, Erdoğan mentioned a grossly exaggerated figure of 170,000 Armenian residents in Turkey. According to him, some 100,000 of these would be illegal migrants. Data from an opinion poll carried out by MetroPOLL among Turkish citizens were discussed. The data showed that 48.8% of the respondents supported Erdoğan's threat to deport the illegal Armenian migrants. The figure coincides with the votes received by the AKP party during the last general elections.[4] According to the Armenian media coverage, Erdoğan's statement was opposed by a large part of the Turkish opposition (*Aravot*, 18 March 2010).

The issue of migrants from Armenia, their possible deportation by the Turkish authorities and the alleged figures was at length discussed in the Armenian and the Turkish media[5]. In the Turkish mass media articles appeared saying that the number of 100,000 was exaggerated and that in reality there were not more than 30-40 thousand migrants from Armenia in Turkey. The head of an Armenian Church parish board in Istanbul, Petros Shirinoghlu, met with Erdogan and stated that the mistake in the speech of the prime-minister happened by Shirinoglu's fault since he once mentioned the figures of 70,000 Armenian nationals in Turkey and another 30,000 migrants from Armenia. On March 27th, during a special program of the "Yerevan" TV channel, the head of the *Migration Agency* of the Republic of Armenia Gagik Yeganyan explained that according to his calculations about 70,000 Armenians left for Turkey during the years 2000-2009. Of these, 60,000 had returned, i.e. the balance was not more than 10 thousand. Yeghanyan mentioned also that in 2004 Turkey's Foreign Minister Abdullah Gül (now president of Turkey) spoke of 40,000 migrants from Armenia in Turkey.

Yeghanyan's estimates were confirmed by empirical surveys. According to the 2009 survey "Identifying the state of Armenian migrants to Turkey: Sociological Qualitative Research"[6] funded by the *Eurasia Partnership Foundation* and conducted among migrants from Armenia in Turkey, the overwhelming ma-

4 For the results of this poll see http://www.metropoll.com.tr/report/oscar-odulleri-farkindaligi-ve-ermeni-sorunu-mart-2010 (in Turkish); comments by Bekdil and Burak: We are all deeply moved! "Hürriyet Daily News", 23 March 2010. http://www.hurriyetdailynews.com/n.php?n=we-are-all-deeply-moved-2010-03-23 (Last accessed 28 June 2010)

5 See for example the critical comment by Burak Bekdil: The Exodus – Part II? "Hürriyet Daily News", 18 March 2010. - http://www.hurriyetdailynews.com/n.php?n=the-exodus-8211-part-ii-2010-03-18 (Last accessed 28 June 2010)

6 For the full report of this survey, see the research paper of project leaders Alin Rozinian: http://epfound.am/files/epf_migration_report_feb_2010_final_march_5_1.pdf

jority of them were women (96%). Half of them (48%) come from the Shirak region of Armenia that is bordering Turkey and is still suffering from the social and economic consequences of the 1988 earthquake. Most women from Armenia seem to have Armenian employers of Turkish nationality. Some 72 % of the Armenian women work as housemaids, 18% take care for sick persons, 6% are employed as shop assistants, and the rest of 4% mentioned jobs in the service sphere. Their situation in Turkey is complicated by the circumstance that Turkey refuses to enter diplomatic relations with Armenia. As a result, there are 12-13,000 illegal migrants of Armenian nationality residing in Turkey.[7] The survey also revealed that according to official data during the last years 600-800 Armenian children were born in Turkey who do not have Armenian or Turkish citizenship and remain stateless. They have the right to attend Armenian classes in Istanbul, where the vast majority of Armenian migrants resides. The future of these stateless children is uncertain.

Periodically problems of trafficking migrants to Turkey and of child trafficking were discussed. A survey funded by the European Union and the association of social workers *Harmonious Society* conducted a survey among 1,200 households in Armenia. The survey aimed at finding out the degree of awareness of child trafficking among the population. The survey revealed that 86.7% of the respondents were informed about this problem (*Aravot*, 31 March 2010).

Information about the activities of the organization *Soldier's Mother* was discussed in the media. Since 1997 the organization has conducted projects aimed to the solution of migration problems connected with the national service. The last project "Civil Society and Migration" has been conducted since 2007 with the assistance of the *Partnership Eurasia Foundation* in seven migration centers located in Armavir, Goris, Sisian, Noyemberyan, Martuni, Gyumri and Vanadzor. The aim was is to bring legal and informational support to out-migrants and also to potential migrants who are going to leave for CIS countries. Special attention is being paid to persons of military age and their families. As part of the project, a consulting center has been established, where over one thousand persons called. The center is processing anonymous electronic correspondence and provides help in more than hundred calls daily. People call for help from Russia, Belarus, Ukraine, Kazakhstan, France, Ireland, Holland, and many other countries.[8]

---

7 Kurt, Süleyman: Report: 12,000 Armenian citizens working illegally in Turkey. "Today's Zaman", 5 December 2009. - http://www.todayszaman.com/tz-web/news-194672-100-report-12000-armenian-citizens-working-illegally-in-turkey.html (Last accessed 28 June 2010)

8 According to the website of the Armenian NGO "Zinvori mayr" ("Soldier's Mother") - www.zinvori-mair-ngo.am

6 Survey on Returnees from Migration and on Potential Migrants

Three different questionnaires were designed and applied in surveys among migrants who returned to Armenia and among those who intended to go abroad but did non have an international migration experience yet. The first questionnaire was used for interviews with 200 residents of Armenia who returned from Moscow after at least one month of stay there. The second one was applied on 300 returnees from all other destinations than the Russian Federation after a period of at least one month of stay abroad. Finally, the third questionnaire was applied on 100 residents of Armenia who did not have an international migration experience but intended to leave abroad in the nearest future to stay there for more than a month. It was impossible to conduct such a survey by using the method of random selection. Instead the 'snow ball' method was applied.

Examining the socio-demographic characteristics of the interviewed samples we find similarities on several parameters but also essential differences. Among the returnees from Moscow the prevalence of male respondents is noticeable. The proportion of men leaving for Russia (and particularly to Moscow) in search of work is undoubtedly higher than that of women. This result is confirmed also by the data of previous surveys. "Approximately three-quarters of the Armenian emigrants in the last decades have settled in the former Soviet Union countries, mainly in the Russian Federation, 15 per cent in various European countries, and 10 per cent in the United States. More than 60 per cent of emigrants are men, of working and reproductive age (20-44), and with average educational level that significantly exceeds the average national standards" (Migration in Armenia 2008: 12; Human Development Report Armenia 1999: 46). Another source informs that "in the 2005 survey 14 percent of men were reported as not living at home at the time of the survey. It might be taken for given that three-quarters of Armenian men who were not living with their family were residing in Russia. The majority (52 percent) of men who were staying elsewhere at the time of the 2005 survey have occupations as skilled manual workers and 7 percent do unskilled manual work. The 14 percent of migrant men did professional, technical, or managerial work" (Johnson 2007: 7).

Table 1: Gender of interviewed migrants (in %)

ArGeMi Cohorts	Male	Female	Total
Returnees from Moscow	67.5	32.5	100.0
Returnees from other destinations	53.0	47.0	100.0

The interview showed that the 300 returnees from "other destinations" came back from 38 countries in North and South America, Europe, Asia and Africa.

Table 2: Migration destination

Country of most recent stay abroad (with duration longer than one month)	**Frequency**	**Percent**
USA	42.0	14.0
Ukraine	29.0	9.7
Georgia	25.0	8.3
France	24.0	8.0
Germany	22.0	7.3
Turkey	17.0	5.7
Other 32 countries	141.0	47.0
Total	300.0	100.0

The above distribution of returnees corresponds with the countries of the largest Armenian Diasporas besides Russia. The snow-ball method applied in the ArGeMi project revealed an authentic image of out-migration from Armenia. There were large Armenian communities in Georgia as well as in Azerbaijan until the end of the 20th century. At present there are nearly no Armenians left in Azerbaijan. Although Georgia suffers from the same economic and social hardships as Armenia, migrants from Armenia go there to sell and buy farm produce. Because of tense relations in Georgian regions with predominantly Armenian population, particularly in Javakhk, many Armenian families migrate to Armenia or send their children to Yerevan to study there. This causes the exchange flows of seasonal and temporary migration between Armenia and Georgia.

Turkey does not maintain diplomatic relations with neighboring Armenia. In 1993 its state border to Armenia was closed for the purpose of supporting Azerbaijan that was at war with Nagorno-Karabakh. Since then the blockade of Armenia has not been lifted. Nevertheless, thousands of Armenians travel to Turkey by bus via Georgia, and businessmen from Turkey come to Yerevan for commercial purposes. Business activities between Armenia and Turkey are growing. The import from Armenia was officially registered in Turkey since 2005 and started to increase since 2008.[9] In January 2010, 92,824 persons left Armenia for Turkey and 89,066 people arrived from Turkey. Half of them returned by buses, the other half by air transport.[10] The balance is 3,758 people, who probably stayed on in Turkey in January 2010.

The following Table shows the distribution of age groups of the ArGeMi respondents:

9 See www.turkstat.gov.tr

10 According to the official site of the *RA Migration Agency* - www.backtoarmenia.am

Table 3: Age of respondents (%)

Age	Returnees from Moscow	Returnees from other destinations	Potential Migrants
18-24	9.0	7.4	8.0
25-34	31.5	23.9	20.0
35-54	46.0	50.6	53.0
55-64 and more	13.5	18.1	19.0
Total	100.0	100.0	100.0

The age group of 25-34 years (31.5%) is the largest among the returnees from Moscow, compared to returnees from other destinations (23.9%) or potential migrants (20.0%). The explanation sounds that migrants in Moscow are mainly occupied in constructions, transportations and trade. There is a need for physically able young labor force.

The educational level among the three interviewed groups is nearly the same and quite high. The level of higher education among the returnees from other destinations than Moscow is 35.7%, while the ratio among returnees from Moscow is 29%. This testifies the assumption that the migrants from Armenia to Russia are mainly labor and office workers with somewhat lower educational level.

Table 4: Migrants with higher education (%)

ArGeMi Cohorts	Total	Men	Women
Returnees from Moscow	29.0	58.6	41.4
Returnees from other destinations	35.7	49.5	50.5
Potential migrants	22.0	45.6	54.5

Table 5 reveals some reverse gender segregation of migrants with higher education. The amount of male respondents with higher education among the returnees from Moscow is larger (58.6%) than among migrants from other countries (49.5%) and larger than among the interviewed potential migrants in Armenia (45.5%). On the other hand, the group with higher education was larger among the interviewed potential female migrants (54.5%) than among female migrants from other destinations (50.5%) and female returnees from Moscow (41.4%).

A difference in the command of foreign languages between the three samples was observed. Particularly the command of the English language differs substantially. Among the returnees from other countries, 34.7% claimed to have an excellent or good command of English. This applied to 23% of the returnees from Moscow. Some 30% of the potential migrants mentioned a good command of English. This might reflect their opinion about their language skills. Among the 300 interviewees who have returned from other destinations than Russia the assessment of foreign language skills (in particular English) looks more reliable due to their experience of living and communicating abroad.

Half of the interviewed migrants proved to be permanent residents of the capital Yerevan. The rest were residents of two big cities and of two smaller cities of Armenia. The employment rate in the provinces (*marzer*) of Shirak, Lori and Syunik is rather low (Minasyan et al. 2007: 19). This is the reason why we chose for the ArGeMi survey two provincial capital cities (Gyumri, Vanadzor) together with two other towns of the Syunik and Lori provinces (Goris and Alaverdi). More than the half of all interviewees from Armenia abroad (56.4%) came from the Yerevan[11], Shirak and Lori provinces. The largest amount of migrants in 2007 was 33,300 migrants from Yerevan, 31,000 migrants from the Shirak province, and 21,500 from the Lori province (Minasyan et al. 2007: 26).

Table 5: Social demographic structure of migrants (%)

City	Returnees from Moscow					Returnees form other countries				
	Gender		Education			Gender		Education		
	Men	Wo-men	Se-con Dary	Profes - sional	Hi-gher	Men	Wo-men	Se-con dary	Profes - sional	Hi-gher
Yerevan	64.0	36.0	21.0	31.0	48.0	49.7	50.3	13.9	31.8	54.3
Gyumri, Vanadzor, Alaverdi, Goris	71.0	29.0	41.0	33.0	26.0	56.4	43.6	36.3	30.8	32.9

Two thirds of the interviewees were people with families (63-65%). The ratio of singles, widowed and divorced in all three interviewed groups was slightly less than a quarter (23-25%). This again indicates that the motif to support the families prevails among migrants from Armenia.

The majority of the ArGeMi respondents have been abroad several times since the early 1990s. About 60% of the interviewed returnees from Moscow and other destinations were abroad 1-2 times abroad; 22-27% of the respondents were abroad 3-4 times and the rest (13-16%) even more often than 4 times. Among the returnees from Moscow the majority (80%) has visited Moscow 1-2 times and the rest of them (20%) has visited that city 3, 4 and more times. Nearly half of the interviewed (47%) made their last trip to Moscow between 2006 and 2008. The remaining respondents made their last trip between 1990 and 2005. On average, the stay in Moscow lasted from one month to a year, as 79.5% of all returnees from Moscow and 76% of the returnees from other destinations report:

11 Yerevan is one of eleven *marzer* ('provinces'), which is the largest administrative unit of Armenia.

Table 6: Duration of stay abroad (%)

Duration of last stay abroad	In Moscow	In other countries
From 1 to 6 months	46.0	56.7
From 6 to 12 months	33.5	19.3
From 1 to 2 years	14.0	17.0
From 2 to 4 years	3.5	5.7
More than 4 years	3.0	1.3
Total	100.0	100.0

The above ArGeMi data correspond with results of previous surveys: "As for the duration of the last trip, the majority of migrants stayed abroad for 5-10 months (68.9%) and the mean duration of the trips coincided with the planned duration of 8 months. Over 75% of the migrants left Armenia by the end of spring, March being the most active month for departures. The majority of migrants (78%) returned to Armenia between October and December, but mostly closer to the New Year" (Minasyan et al. 2007: 29). Every fifth respondent gave as reason for the return from Moscow the end of the work contract or the end of studies at the universities. The vast majority of respondents, however, have returned because of personal reasons. Among the returnees from other destinations the personal reasons numbered less than a third, while the percentage of non-personal reasons for a return is higher.

Table 7: Reasons for return (%)

Returnees	Personal reasons	End of contract or studies	Visa expiry	Deportation	Total
From Moscow	63.0	19.5	7.0	3.5	100.0
From other destinations	30.7	24.7	35.0	4.0	100.0

The main reasons why Armenian migrants return from Moscow are personal or family motives since visa is not required for a trip to Moscow and the travel expenses are cheaper than the travel expenses to countries outside the CIS. The migrants to Russia (and particularly to Moscow) can freely move and repeatedly return for their personal affairs, but corruption offers 'chances' as well. The 2007 OSCE survey describes the ambiguous situation as follows: "For most migrants working in Russia, the biggest problem is the police, who check the migrants' documents every now and then. Since a majority of the migrants do not have proper documents, the police often let them go only after paying a bribe. The migrants are afraid of encounters with the police not only because they mean additional expenses but also because the migrants feel they may be subjected to bad treatment, including physical violence. On the other hand, especially for those who leave to work abroad often, it is important to avoid deportation, because they can lose their main source of income. However, the migrants see deportation as quite an unlikely outcome of an encounter with the police" (Minasyan et al. 2007: 50).

The situation of those who have returned from other destinations is different. A visa or work permission is required in order to enter a country of the Schengen Agreement or the so-called «far abroad». The travel expenses are higher, and for this reason most migrants avoid to return, but stay on until the visa expires, the working contract or studies ends, or until the authorities deport them.

In general, it is important for migrants to legalize their stay in the countries of destination. In both samples of returnees from Moscow and from other destinations the most frequently reported difficulty was the legalization of the stay abroad (24.5% and 18.7%). Not less difficult for the migrants was to find a job (10.5% and 16%).

Table 8: Main difficulties abroad (%)

Returnees	Legalization	Employment
From Moscow	24.5	10.5
From other destinations	18.7	16.0

Besides the main problems of legalization of the stay and employment the returnees from Moscow reported about many other difficulties connected with health care (10%), receiving information about their rights and duties in the country of stay (6%), discriminatory attitude of the local authorities (5.5%), development of their own business (4.5%) and housing (3.5%). The returnees from other destinations also indicated a multitude of difficulties related to health care (5.3%), receiving information about their rights and duties in the country of stay (2.3%) and discriminatory attitude of the local authorities (0.7%). The representatives of this sample emphasized the problems connected with the development of their own business (6%) and housing (5.7%).

It is rather symptomatic for the situation of migrants from Armenia that in case of a problem the interviewees would apply for help to the local citizens whom they already know (32.5% and 25.3%), rather than to the local authorities or the Armenian embassy in that country. In other words, in case of need, representatives of both samples apply either to their compatriots or to the local acquaintances (18% of the returnees from Moscow and 13.3% of returnees from other destinations), but not to the official bodies or human rights organizations and not even to the Armenian Church there. According to a 2005 sociological survey on trafficking and labor exploitation of Armenian migrants, only 9.1% of migrants have sought the help of an Armenian representation abroad. The vast majority of them (87.9%) do not even know any organization or NGO abroad that would help persons who suffered violence (Poghosyan 2005b: 46-47). After conducting this survey in 2005, we gave recommendations to widely inform Armenian migrants about the local organizations, NGOs and diplomatic representations of Armenia to which the migrants could apply for help abroad if necessary. After these recommendations and with the support of the *State Agency on Migration* and the IOM, a large quantity of informative booklets were prepared, published, and sent to all Armenian embassies abroad. These data were also published on the website of the *Agency on Migration.* In spite of these activities, the share of those migrants who apply for help has not sharply increased. Evidently, there is a profound distrust of migrants to the state bodies. Migrants are aware of the general disapproving attitude of the officials of the Republic of Armenia to persons who leave the country. May be this is the major reason why migrants don not apply for help to the Armenian embassies.

Table 9: Help appeal (%)

Returnees	To the local acquaintances	To the compatriots
From Moscow	32.5	18.0
From other destinations	25.3	13.3

Referring to the above answers, one may understand why the vast majority of migrants from Armenia felt unprotected during their last stay both in Moscow (74%) and in other destinations (89.3%). The returnees from Moscow mentioned more reasons for feeling unprotected (26%) than migrants from other countries (10.7%). As explanation for the xenophobia in Moscow the returnees from Moscow mentioned the general antipathy against foreigners in the first place ("Muscovites do not like foreigners" - 14.5%) and migrantophobia ("Muscovites do not like labor migrants" - 10.5%). 15.5% of the respondents felt threatened or in danger during their last stay in Moscow, while 19.5% mentioned that they were verbally offended. In other countries the ratio is 1% of those respondents who felt threatened or in danger, while 13% were verbally offended. Compared with the cohort of returnees from Moscow, these rates are considerably lower. In conclusion, migrants leave Armenia not in search of a good life. The majority of the returnees from Moscow experienced persecution, humiliation and exploitation. The attitude towards migrants from Armenia in other countries was more neutral. At least they did not feel threat or serious danger during their stay in these countries, but the majority of them felt nevertheless unprotected.

The majority of returnees from Moscow (65%) were married or lived with a partner. A share of 63.5% have children, 44.5% have 2 or 3 children. Among the returnees from other destinations, 63% were married or were living with a partner and 71.3% have children.

During their stay in Moscow, 89.5% of the migrants had permanent contacts with their family and relatives in Armenia. As to the frequency of contact, 25.5% used to have daily contacts, 54% every week and 10% every month.

Nearly the same results derive from the returnees from other destinations. Some 89% of them used to have permanent contacts with their family and relatives in Armenia. For 20.7% the contacts have been on daily, for 50% on weekly and for 16.7% on monthly basis. Thus, the returnees from other destinations used to have slightly less frequent contacts with their families in Armenia.

All the difficulties like threats and dangers that migrants from Armenia experienced abroad notwithstanding, a considerable part of them agreed with the statement that migration from Armenia is blessing for the Armenian people because they can travel and work abroad freely. But the majority of the interviewed migrants disagreed with this assumption.

Table 10: Migration is a blessing for Armenians ("Yes" answers, in %)

ArGeMi Cohorts	Total	Men	Women
Returnees from Moscow	46.3	45.2	47.7
Returnees from other destinations	43.6	42.2	45.4
Potential migrants	30.1	30.0	30.1

Migration is considered as a blessing by those respondents who already have a migration experience. The interviewed men and women closely agree in this issue. At the same time the overwhelming majority of the interviewees supported the statement that the emigration from Armenia is a curse for Armenians because migrants suffer exploitation and discrimination abroad, and they families suffer as well. On the basis of the ArGeMi polls, we may assume that migrants that migration is a 'blessing' for Armenians because there is a lack of job vacancies in the country and the high level of unemployment forces migrants to search for employment abroad. Due to the emigration of surplus labor, the social tensions within Armenia are eased or prevented. But on the other hand, migration is a 'curse' for Armenians since migrants become a risk group abroad. Some of them become victims of trafficking.

With regard to Armenia, most interviewees see out-migration as a curse for the country because it loses its best people. More men than women agree with this statement:

Table 11: Out-migration is a curse for Armenia ("Yes" answers, in %)

ArGeMi Cohorts	Total	Men	Women
Returnees from Moscow	82.0	85.3	80.1
Returnees from other destinations	84.7	89.2	84.5
Potential migrants	85.0	91.3	74.0

More than a third of the returnees from Moscow believed that some day migrants from Armenia would return home, while a larger part of the interviewees did not believe in this. For the returnees from all other destinations, the level of optimism was even lower. To the contrary, the optimism was higher among the potential migrants.

Table 12: Will the migrants from Armenia return? (%)

ArGeMi Cohorts	They will return	They will not return
Returnees from Moscow	35.5	48.5
Returnees from other destinations	24.7	55.0
Potential migrants	44.0	42.0

Those migrants who had been abroad in non-CIS countries least of all believed that the migrants from Armenia will ever return. Most probably these respondents did not believe in the improvement of the situation in the country. They see no reason to leave what they have achieved abroad and to return to Armenia after the years spent in their difficult integration into foreign societies. Whatever a prosperous country Armenia would eventually become, it could not compete with Europe in terms of the level and quality of living conditions in the foreseeable future. In the interviews with the experts the doubts about the probability of return were articulated in details. In this issue the opinions of experts and of interviewed migrants fully coincided.

Table 13: Will migrants go abroad? (%)

ArGeMi Cohorts	Intend to go abroad
Returnees from Moscow to Moscow	24.0
Returnees from Moscow to other destinations	19.3
Potential migrants to Russia	46.0
Potential migrants to other destinations	51.0

A quarter of the returnees from Moscow intends to re-migrate to Moscow (basically to get a job there), while a fifth intend to re-migrate but to other destinations (again basically to get a job). Thus, among the returnees there are a lot who are ready to leave the country again. But there are also returnees from Moscow or elsewhere who have probably failed abroad or have not managed to settle properly. There are not so many among them who want to leave again. But there are also a lot of young people in Armenia who do not have migration experience and are inclined to leave the country. They still cannot imagine the difficulties the migrants, especially the illegal ones, may face abroad. So it is very appropriate to provide the potential migrants with as much information as available concerning the migration laws, rules and the living conditions in the countries of their destination. Probably some of the potential migrants will change their mind and cancel their trip or at least will get better prepared before their departure. In any case, reliable knowledge about the conditions for migration in different countries will help migrants to better adapt their expectations and plans to realities. During the recent years a lot has been done in this area. Special information materials were published in the mass media. Useful information might be found on official internet sites, particularly on the site of the *State Agency on Migration* and the *Ministry of Diaspora of the Republic of Armenia.*

7 Remittances

Currently the global amount of remittances is at a historically high level. According to estimations the remittances to the developing countries made out US$ 308 billion in 2008 and US$ 293 billion in 2009 (Human Development Report 2009: 72). The *World Bank* defines the "migrant remittances" as the sum of workers' remittances, compensation of employees, and migrants' transfers. "Workers' remittances", as defined in the IMF Balance of Payments Manual, are the current private transfers from migrant workers who are considered residents of the host country to recipients in their country of origin. If the migrants live in the host country for a year or longer, they are considered residents, regardless of their immigration status.[12]

The *Central Bank of Armenia* informs that the remittances from the worldwide Armenian Diaspora are double the size of the country's budget and keep many families above the poverty line. According to a study of the *International Monetary Fund* (IMF), most of the remittances sent to Armenia come from the Russian Federation (70%), and correlate strongly with Russia's GDP growth. The study also indicates that Armenians in most cases send remittances in US dollars (90%) and through banks (from 55 to 85%) (Oomes 2007: 19). According to the IOM, the total amount of incoming migrant remittances into Armenia was 1,273 million USD for 2007 (Migration in Armenia 2008: 19).

During the period between 2005 and 2007 the labor migration affected about 15% of all Armenian households. Approximately 47% of all the households with labor out-migrants would belong to the group of the very poor if they would not receive remittances from their migrant family members. If the monetary support provided by labor migrants to their families would be deducted from the income basket of the Armenian households, the number of poor in the country would increase by 20%. In order to keep poverty at the same level, a much larger state budget than the current one would be required. The conclusion is obvious: Labor migration significantly alleviates the economic burden of a large number of Armenian households and reduces the poverty level.

The report "Remittances in Armenia: Size, Impacts, and Measures to Enhance Their Contribution to Development" shows that remittances contributed to the reduction of poverty in Armenia (Roberts 2004: 61). The 2008 Armenian household survey conducted by the Armenian National Statistical Service (NSS)

---

12 Cf. http://siteresources.worldbank.org/ EXTDECPROSPECTS/ Resources/476882-1157133580628/FactbookDataNotes.pdf)

revealed that the remittances contribute to the reduction of inequality. According to income data reported in the survey, remittances make up to 80% of the income of the receiving households. Moreover, remittances appear to be directed to some of the most vulnerable households in Armenia. The survey also highlighted the point that there is nearly the same percentage of the urban and rural households that receive remittances. However, the rural households receive relatively more remittances from the CIS countries and less from the United States and Canada (Migration in Armenia 2008: 21). The remittance flows from Russia to migrants' countries of origin have significantly grown as compared with the inward flows sent by Russians from abroad (Migration in the Russian Federation 2009: 41).

The volume of remittances made out $2.5 billion in 2008 and made out 20% of the GDP of Armenia. Almost 80% of it came from Russia. The remittances decreased by 25% in 2009 (Grinyaev 2009: 16-17). Remittances have been particularly important for Armenia during the transition period. They were a crucial financing component that enabled Armenia to run large deficits with the outside world and still maintain decent living standards and investment. Remittances continue to be an important source of external deficit financing. The positive or negative shocks to remittance flows have repercussions for the Armenian macro-economic situation (IOM 2008: 20). The introduction of two transfer systems in the Commonwealth of Independent States (CIS), *Anelik* and *Unistream* significantly lowered transactions costs and were thus encouraging remittance transmission through formal and controllable channels. The returnees largely preferred bank transfers for sending money to their families. According to the ArGeMi poll 77% of the returnees from Moscow were sending money by bank transfers, while only 23% were sending the money in cash through carriers. Nearly the same percentage of the migrants to other destinations (75.3%) used to send their remittances by bank transfers, while 24.7% preferred personal carriers.

In the ArGeMi project a considerable part of the returnees from Moscow answered that they had the opportunity of sending money to their relatives (40.5%), while 32.5% answered that they did send money regularly. Another 8% answered that they sent money home from time to time. The returnees from other destinations had sent money to their relatives to a lesser degree (36.3%). 33% of these had sent money regularly and only 3% answered that they had sent money from time to time. As a rule, the amounts they sent were not less than 100 USD per month, with an average amount of 250 USD per month. The amounts identified in the ArGeMi polls are thus slightly higher than the results obtained by the 2007 OSCE survey. According to the OSCE data, in 2007 the total amount of money sent home by every Armenian labor migrant was 2,720 USD that is 226 USD per month on average (Minasyan et al. 2007: 31):

Table 14: What was the average amount of US$ you used to send to Armenia per month? (%)

Amount (in US$)	Returnees from Moscow	Returnees from other Destinations
Less than 100	7.0	1.7
From 100 to 200	13.0	10.7
From 200 to 300	9.0	10.7
From 300 to 500	7.0	8.7
Over 500	4.5	4.7
No answer	59.5	63.5
Total	100.0	100.0

If we take into consideration that in 2009 the average monthly salary in Armenia was about 150 USD and the minimal salary was about 80-85 USD, the above amounts of remittances were an essential support for the families in Armenia. According to the interviews with returnees from Moscow, 7% of their families in Armenia had a monthly income of less than 100 USD. Among the returnees from other destinations such a very low monthly income had 3.7% of their families.

How are the remittances spent in Armenia? The *International Center for Human Development* made the following general remarks: "It is worth noting that remittances from foreign countries are mostly used for consumption. Another part is directed to payment of educational fees for family members. Actually, the scope of investments in education is growing due to migrant remittances. Many Armenian banks have started to treat the remittances sent by labor migrants as a stable source of income and a sufficient basis for providing consumer loans to the households. Those migrants who have lived abroad for 5-10 years send the lion's share of remittances. In contrast, Armenian citizens living abroad for more than 10 years tend not to send any money at all. Most probably residing in a foreign country for over ten years, they lose touch with their homeland and one of the reasons might be that over this time they take their families and relatives with them" (Policy Brief : 1).

8 Legislation on Migration and Migration Policy in Armenia

In a broad understanding of the term, migration related legislation includes laws, decrees and regulations not only on migrants, but also on refugees and asylum applicants arriving in Armenia. It covers both internal and external migration. Migration related legislation certainly needs improvement in the Republic of Armenia. The existing regulations passed by the National Assembly cover the issues not in an appropriate and complete way. Relevant aspects referring to migration issues are regulated by separate rules and governmental decrees. For example, until 2007 the legal field that regulated migratory problems of foreigners was not coherent and in several cases did not correspond with the internationally accepted and used models. Sometimes these regulations did not even correspond with the current Constitution of Armenia. Frequently the legislators left substantial decisions to the discretion of the executive authorities. By doing so they put into question the democratic principles of law enforcement practice in migration issues (Kabeleova, Mazmanyan and Yeremyan 2007: 27).

After proclaiming its independence, the Republic of Armenia adopted a number of laws that were intended to regulate the issues pertaining to migration. Among the major laws adopted were the following:

- The *Law on the Legal Status of Foreign Citizens in the Republic of Armenia* – 1994
- The *Law on Citizenship* - 1995
- The *Law on IDPs* – 2000
- The *Law on Political Asylum* – 2001
- The *Law on Refugees* – 1999 (amended in 2002)
- Draft *Law on Labor migration.*

A serious problem of the Armenian legislation on migration issues concerns its democratic nature. The projects of laws were discussed several times and were reviewed in different committees of the Parliament. However, the government resolutions and decrees, and also other governmental acts like the rules established by the state ministries and agencies are not being reviewed thoroughly. The working of those bodies lacks transparency. The imprecise distribution of responsibilities among different administrative bodies promotes overlapping functions and imprecise execution of laws. "Sub-legislation acts, as a rule, hamper the formation of a stable and effective legal field and administrative limits. Such an approach may be effective for a short period of time but nevertheless it must be realized that a clear and distinct legislation must be worked out, and a proper democratic control must be established" (Kabeleova, Mazmanyan and Yeremyan 2007: 46).

In June 2004 the government of Armenia adopted a concept on population migration. It contains the principles of the state migration policy, priorities, and possible mechanisms and directions for their solution. It also outlines the changes required for a harmonization of the legislation with the priorities of the state migration policy. The concept lists the state bodies engaged in the management of migration issues and their functions, the separation of their functions, and the mechanisms securing active collaboration between the migration system bodies. These and a number of other actions allow an improvement of Armenia's migration legislation and the regulation of migration processes. However, it is obvious that along with the improvement of the legislative basis the control over the law enforcement practice must be reinforced and the administrative structures must be strengthened.

From the beginning of 1989 a *State Committee on Migrants and Refugees* operates in Armenia. It was created to provide help and work for the 360,000

Armenian refugees from Azerbaijan. Over the time it expanded its activity to migrants and asylum seekers from different countries. However, the assistance and impact of social and international organizations in this field is hard to overestimate. Today, in Armenia there is a net of 10-15 NGOs that work on migration issues. Many of them have accumulated profound competence through their work with refugees, illegal and transit migrants, victims of trafficking, and migrants returning to Armenia.

The current system of statistics on out-migration is based on data of the population registration and on the registration carried out by the passport services of the territorial divisions of the Police of the Republic of Armenia. This statistical coverage does not reflect the real volume of population movements. In its current form, statistics on the population migration are more or less identical with the statistics on registered population. But a considerable number of persons leaves the country and resides abroad for a rather long period of time without being erased from the register. In order to acquire a more precise understanding of the external migration volumes, the *State Department for Migration and Refugees* in co-operation with the *National Statistical Service*, transportation companies and services of the border-crossing posts of the Republic of Armenia have initiated a data collection on arrivals and departures from Armenia. Since 2000 the Department monthly receives corresponding information

- From the Central Department of Civil Aviation,
- From the State Customs Committee,
- From the Border Guards of the National Security Service.

According to data of the *Agency on Migration of the Ministry of Territorial Administration of the Republic of Armenia*, the quantity of arrivals and departures in March 2008 was the following:

Table 15: Number of arrivals and departures of persons (March 2008)

	Total	**Including transport**		
		Airway	Railway	Car
Arrivals	88,524	38,423	1,137	48,964
Departures	104,109	56,369	1,301	46,439
Balance	-15,585	-17,946	-164	2,525

The analysis of the data on passenger registration made at the entry points of Armenia shows that in the period from January to March 2008 the quantity of departures outnumbered the quantity of arrivals by 32.1 thousand persons. This is 10.9 thousand persons more than during the same period in 2007. The negative balance of 15.6 thousand in March 2008 alone in comparison with March 2007 showed an increase of 5,4 thousand persons. In March 2009 the quantity of arrivals and departures 2009 was as follows:

Table 16: Number of arrivals and departures of persons (March 2009)

	Total	Including transport		
		Airway	Railway	Car
Arrivals	85,619	37,654	928	47,037
Departures	88,671	44,084	955	43,632
Balance	-3,052	-6,430	-27	3,405

During the period of January-March 2009 the number of departures still outnumbered the arrivals by 10,947 thousand persons. However, the difference was three times less than it was in 2008. The negative balance in March 2009 alone was 3,1 thousand persons which in comparison with March 2008 have decreased three times.[13] According to the data of the *Migration Agency* in March 2010 the total number of arrivals was 111,098 and the number of departures 104,744. For the first time the balance in March 2010 was positive – 6,354 persons.[14]

The migration flows intensify in times of political instability. This happened, for example, in 2003 during the USA war in Iraq. At that time several thousand refugees from Iraq – most of them ethnic Armenians - arrived in Armenia. The majority of these immigrants were Iraqi citizens. In other cases larger return migrant flows come about. During the military operations of Georgia against South Ossetia and the reaction of the military forces of the Russian Federation, about 20 thousand migrants returned from Georgia to Armenia. There were 8-10 thousand foreigners among them who left Armenia to Europe and USA. There were also ethnic Georgians and Armenians of Georgian citizenship among them who left their homes after martial law was imposed over Georgia. Because of the termination of the aviation transport between Moscow and Tbilisi, Georgians who lived in Russia were returning to their homeland via Armenia.

Recently, the *Migration Agency of the Ministry of Territorial Administration* elaborated a concept on the return and reintegration of migrants in Armenia. In the recent years a slow but stable return flow of Armenian citizens from foreign countries can be observed. Part of these migrants had no legal status in the destination countries. In order to legitimate their status they apply for asylum, but as a rule their applications are rejected. They face the options of return or deportation to Armenia. As an integral part of the *Neighborhood Policy* of the European Union Armenia had signed readmission agreements with Switzerland, Denmark, Germany, Lithuania and Bulgaria in order to take back Armenian citizens who reside in these countries without legal ground. Negotiations on readmission agreements with other countries continue. After their return to Armenia some migrants depart again from the country because they find themselves in the following situations after having spent years abroad:

13 Cf. http://backtoarmenia.am/upfiles/2008balance.html
14 Cf. http://backtoarmenia.am/upfiles/2008balance.html

- A totally changed legal field and new institutions;
- disappearance of social connections of relatives, kin, friends;
- stressful situation connected with the departure from the home country and then with the return to it (Poghosyan 2006b)

Thus, a returnee finds him or herself in a new environment and usually looks for ways to go back abroad. The survey among returnees conducted by the OSCE office in Armenia confirmed that the majority of returnees do not get proper job offers in Armenia. *"Some people, after failing to find appropriate employment following their return with the family, had to leave the family in Armenia and migrate for work abroad"* (Minasyan et al. 2008: 35). In order to avoid re-migration some provisions for reintegration assistance to returnees seem indispensible. The implementation of the *Return Assistance Program for Nationals of the Republic of Armenia from Switzerland* conducted since 2004 by efforts of the *Federal Office for Migration* (FOM) of Switzerland, the Swiss Development and Cooperation Agency as well as the *Migration Agency of Ministry of Territorial Administration of the Republic of Armenia* confirms this point. As a result of this program no returnee from Switzerland has ever since departed from Armenia within more than four years.

9 Conclusions and Recommendations

The main flow of cross-boundary migrants is directed towards Western Europe and North America. After the collapse of the Soviet Union, many thousand migrants from he post-Soviet successor states joined the international migration flows. Together with the external migration, the population transfers inside and between the countries of the former Soviet Union increased considerately. According to the UNHCR, in this territory the large number of 2.3 million internally displaced persons was registered in 1991-1992. The main migration trend on the post-Soviet space was the movement from or to Russia. Russia is the main destination for the majority of migrants from the CIS countries.

The South Caucasus changes its traditional image. The main changes during the last decade of 20th century were the depopulation of the area as a result of massive emigration. The historically unprecedented process of depopulation was but one of the many negative consequences of the collapsed Soviet system. Altogether about three million people left the South Caucasus, including one million people from Armenia. The main reason for the massive outflow of population was the abrupt reduction of work places, the deterioration of the living conditions and currently the international financial and economical crisis. Armenia became an exporter of labor force and in particular of skilled workers. The export of the labor resources and the outflow of the economically active population on a large scale essentially changed the demographic structure of Armenian society.

The social-demographic structure of migrants from Armenia includes people from quite different strata of the Armenian society. But mostly the healthy, young, and professionally skilled citizens departed. In general they are in the working age. It is destructive for a small country to spend means on preparing specialists and then export them. With «brain drain», Armenia loses the intellectual and spiritual wealth of the country that has been developed for decades. It will be rather difficult to replace these human recourses. The experts confirm that also in the future Armenia will face the excess of labor resources. Subsequently, the people will continue to search for a job out of Armenia that will continue the outflow to other countries. The 2010 UNDP national migration report on Armenia stated that 200-300 thousand Armenians would leave the country in the nearest future (Human Development Report Armenia 2010).

The key problem of the migration legislation in Armenia is that many procedures are not clearly defined in the laws and sub-legislative acts, including the resolutions and decrees of the government as well as many rules and orders established by state committees and commissions. This legal uncertainty has implications. The common answer of the majority of interviewed during the ArGeMi survey of experts was that none of them possessed exact information on the amount of migrants, particularly to Russia and in Moscow. The experts deplored the lack of information and doubted its reliability as far as migrants from Armenia to Moscow were concerned. An estimated 95 percent of the total number of migrants from Armenia to Russia seeks jobs there. Since the autumn of 2008, Russia was affected by the deepening global economic crisis. It resulted in the dwindling of real estate construction and other sectors that used to provide jobs for foreign workers.

Consequently, at the turn of 2008 a reverse migration trend was observed. A "positive migration balance" of 1,200 people was reported for March of 2009, with nearly 70,000 people who stayed on instead of departing. Most probably, they joined the army of unemployed in Armenia. The official statistics for the January-May 2009 period showed a decline of roughly 40 percent in remittances against the same period in 2008. In January 2009, remittances decreased by 25 percent and in April 2009 by 39 percent. The same tendency continued in the first three months of 2010, when remittances decreased by 17 percent. In 2009-2010 the government of the Russian Federation reduced the labor quotas for migrants from the CIS countries. Many Armenians remained unemployed in Russia and had to return. The economic crisis in Russia reached its peak in winter 2009 when migrants, as a rule, come back to spend the New Year holidays with their families. Indirectly, the impact of the global crisis on Armenia's economy became manifest in the reduction of remittances on one hand and in the growth of the unemployment level on the other hand.

When asked in the ArGeMi polls about their opinion concerning the prospects for return of out-migrants to Armenia, both migrants and experts were not optimistic. They assumed that there is currently no return flow of migrants, because the socio-economic situation in Armenia has not improved enough. Only 35.5% of the interviewed returnees from Moscow believed that some day the migrants will return to Armenia. The returnees from other destinations expressed even stronger doubts about the return potentials. To the contrary, re-migration potentials are quite high among the temporary returnees. A quarter of the returnees from Moscow intend to re-migrate to Moscow to get a job, while 19.3% intend to re-migrate to Western Europe. Among the 300 interviewed returnees "from other destinations", the most frequently mentioned destinations were the USA, Georgia, France, the Czech Republic and Germany. In all, 38 countries of destination were named. According to the results of the ArGeMi survey, the main purpose of migration was the search for jobs abroad. Therefore, the legalization of work and residency were named as highest priorities. In Moscow and in other destinations, returnees found it most difficult to legalize their stay in the host country. Not less difficult for them was to find a job abroad at all.

The interviewed respondents returned home after they had spent a certain time abroad. The reasons for return were different. The main factors of return from Moscow were personal or linked with the family. Migrants who leave for Russia (and for Moscow in particular) repeatedly return to Armenia for taking care of their personal affairs because they do not face a visa regime and the travel expenses are relatively low. The situation is quite different for the returnees from other countries. Travel expenses in these directions are higher. The migrants do not return home until the visa expires, a work contract or studies are terminated, or until they are deported by the authorities.

The ArGeMi respondents reported difficulties connected with the health care, with the access to information about their rights and duties in the country of stay, discriminatory attitude of the local authorities, problems linked with the development of their own business, and difficulties in arranging the housing. In all these cases the migrants would rather address for help compatriots or local citizens whom they know than the local authorities, the Embassy of the Republic of Armenia in the given country or human rights organizations. As a rule, the large majority of migrants from Armenian felt unprotected during their stay both in Moscow and in other destinations.

The ArGeMi survey also revealed that migration is perceived as a blessing mostly by those respondents who already have a migration experience. At the same time the overwhelming majority of the interviewees mentioned that out-migration from Armenia is a course for Armenian nationals because migrants suffer exploitation and discrimination abroad, and they families suffer as well. As for Armenian society, most respondents saw migration as a curse because the country loses its best human capital.

In the course of the last 20 years several laws and governmental decisions related to out-migration were issued. A special migration service was established. There are intergovernmental agreements on readmission, etc. But there are still various open problems connected with the legal regulation of out-migration. The major issue is the massive depopulation with all its negative consequences like the decrease of birth rates, gender imbalance, social structure deformation, and many others. It is evident that migration problems cannot be solved in Armenia or in Russia alone. They are common issues like the ecological problems. Scholars have to raise the awareness of the authorities in the concerned countries. The migration space of the CIS countries is a common one. It was generated by the collapse of Soviet Union. The problems of labor migration can be solved only by common efforts of the receiving (Russia, Kazakhstan, Ukraine), as well as the sending countries (Armenia, Georgia, Tajikistan, Kyrgyzstan, Moldova). A common migration space needs a common regulation, and common rules are requested.

Some 55-60% of the out-migrants from the previous Soviet republics stay within the borders of the CIS. Despite this fact, the governments of the CIS member and associated states never assembled their experts in order to work out a common migration policy. One reason for this striking development could be the conservatism of the authorities concerned and their indifference to the destiny of the people and the countries. This indifference shows also a lack of understanding of the situation and an unskilled management of rather serious problems. While not expecting constructive solutions from the authorities, the migrants themselves made their adaptations in the direction that in the future civilized routes will be developed. This adaptation process began long ago, but it is never late to regulate and legalize it for the benefit of all participants in migration. It should be explicitly taken into account that there is no one single reason for migration. As to labor migration, there are countries with a low level of employment and there are countries with a high demand in labor resources. Hence, migration as redistribution of labor resources takes place. This happens largely unregulated and is therefore connected with losses, exploitation, crime and many other negative consequences. To avoid this, migration processes should be included into the civilized discourse and should be regulated by the involved CIS countries in the framework of the CIS.

In the case of Armenia and Russia the problem is simple. Armenia needs its citizens and cannot afford that they move to Russia to settle there permanently. Russia, on the other hand, needs foreign workers and specialists, including the ones from Armenia. Therefore, we would recommend that Armenian nationals can easily leave for earnings in Russia without particular problems and within an official agreement. In that case one will reside in Armenia and work in Russia. This will benefit both Armenia and Russia, and above all the migrants themselves. Some other suggestions might be formulated as follows:

1. To organize the departure and entry for labor migrants within CIS countries on a clear legal basis with preferences for organized groups of labor migrants;

2. To create a CIS agency for dealing with labor migration;
3. To coordinate and improve the national legislations on cross-boundary migration;
4. To regulate the organized flow of labor migrants by means of quotas;
5. To strengthen the legal protection of labor migrants by stressing the responsibilities of both the countries sending and receiving migrants;
6. To maximally simplify the money transfers by legal channels and increase the security of the transfers.

Thus, the ArGeMi project leads to the conclusion that countries like Armenia might particularly gain from the circular migration by receiving returnees with an increased skill, knowledge and resources due to their stays abroad and particularly in Russia.

References

ARAKELYAN, Vahagn, Manuk HARUTIUNYAN and Gevork POGHOSYAN (2007): *Armenia: Alienated Society*. Yerevan: Institute of Philosophy, Sociology and Law

CASTLES, Stephen (1999): 'International Migration and the Global Agenda: Reflections on the 1998 UN Technical Symposium'. *International Migration*, Vol. 37 (1) March 1999, pp. 5-19

CASTLES, Stephen and Mark MILLER (1996): *The Age of Migration: International Population Movements in the Modern World*. London: Macmillan

DADRIAN, Vahakn (1995): *The History of Armenian genocide: Ethnic Conflict from the Balkans to Anatolia to the Caucasus*. Providence: Berghahn Books

GENOV, Nikolai (2009): 'Labor as Commodity in the Global Market'. In: Poghosyan, Gevork (Ed.): *Armenian Society in Transition*. Yerevan: Armenian Sociological Association, pp. 171-200

GRINYAEV, Sergey (2009): 'Transcaucasia amid the Global Crisis; Resume' *21-rd Dar* [21st century], Yerevan: Noravank Foundation, No. 2, pp. 16-17

HAGOPYAN, Gayane (1988): 'The Immigration of Armenians to USA'. *Armenian Review*, Vol. 41, No. 2/162, pp. 67-76

Hay spyowrk' hanragitaran [Armenian Diaspora; Encyclopaedia] (2003) Yerevan: Hanragitaran (in Armenian)

Hayadaran: Hamahaykakan gid-owġec'owc' 2009 [All Armenian Business Catalogue and Guide-Explorer 2009] (2009) Vol. 1 and 2, Yerevan: Gorcarar Tiezerk (in English, Russian and Armenian)

Human Development Report: Armenia 1998 (1999) Yerevan: UNDP

Human Development Report 2009: Overcoming Barriers: Human Mobility and Development. (2009) New York: Palgrave Mackmillan

Human Development Report: Armenia 2010 (2010) Yerevan: UNDP

IVAKHNYUK, Irina (2006) *Migrations in the CIS Region: Common Problems and Mutual Benefits.* International Symposium "International Migration and Development", June 28-30, 2006, Turin http://www.un.org/esa/population/migration/turin/Turin_Statements/IVAKHNYUK.pdf

JOHNSON, Kiersten (2007): 'Migration, Economy and Policy: Recent Changes in Armenia's Demographic and Health Indicators'. *Armenia Trend Report.* Calverton: Macro International, July 2007 http://www.measuredhs.com/pubs/pdf/TR3/TR3.pdf

KABELEOVA, Hana, Armen MAZMANYAN and Ara YEREMYAN (Ed.s) (2007): *Assessment of the Migration Legislation in the Republic of Armenia.* Yerevan: OSCE

KARAKASHIAN, Meliné and Gevork POGHOSYAN (2003): 'Armenian Migration and a Diaspora: a Way of Life'. In: Adler, Leonore Loeb and Uwe P. Gielen (Ed.s): *Migration: Immigration and Emigration in International Perspective.* Westport: Praeger Publishers, pp. 225-242

KHARATYAN, Hranush (Ed.) (2003): *Artagaxtẹ Hayastanic* [The Emigration from Armenia]. Yerevan (in Armenian)

KHATANESYAN, E. (1965): *Hayoc' tiv* [The Number of Armenians]. Boston: Hairenik Press [in Armenian]

KHOJABEKYAN, Vladimir E. (2004): *Regularities of the Demographic Processes in Armenia in the 19th and 20th centuries and at the Threshold of the 21st Century.* Yerevan: The National Academy of Sciences of the Republic of Armenia, Institute of Economics

KOUTCHARIAN, Gerayer (1989): *Der Siedlungsraum der Armenier unter dem Einfluss der historisch-politischen Ereignisse seit dem Berliner Kongress 1878: Eine politisch-geographische Analyse und Dokumentation.* Berlin: Dietrich Reimer Verlag (Freie Universität Berlin/ Abhandlungen des geographischen Instituts - Anthropogeographie. Bd. 43)

MARCINIAK, Tomasz (2001): 'Armenians in Poland after 1989'. In: Poghosyan, Gevork (Ed.): *Migration in Armenia.* Yerevan, ASA, pp. 107-116

Migration in Armenia: A Country Profile (2008). Geneva: IOM

Migration in the Russian Federation: A Country Profile 2008 (2009). Geneva: IOM

Migration Trends in Eastern Europe and Central Asia: 2001-2002 (2002). Geneva: IOM

MINASYAN, Anna, Alina POGHOSYAN, Tereza HAKOBYAN and Blanka HANCILOVA (2007): *Labor Migration from Armenia in 2005-2007: A Survey.* Yerevan: Asoghik, http://www.osce.org/publications/oy/2007/11/28396_996_en.pdf

MINASYAN, Anna, Alina POGHOSYAN, Lilit GEVORGYAN and Haykanush CHOBANYAN (2008): *Return Migration to Armenia in 2002-2008: A Study.* Yerevan: Asoghik http://www.osce.org/publications/oy/2009/01/35901_1225_en.pdf

OOMES, Nienke (2007): *Coping with Strong Remittances: The Case of Armenia*. Yerevan: International Monetary Fund Representative in Armenia. http://www.imf.org/external/np/seminars/eng/2007/kazakhstan/oomesppt.pdf

PANOSYAN, Razmik (2003): *The Armenian Diaspora Today: Lobby, Political and Identity Groups*. Paper Presented at the World Congress of Basque Centres of Clubs, Victoria-Gasteiz, Spain http://www.lehendakaritza.ejgv.euskadi.net/r48-2312/en/contenidos/informacion/congresos_vascos/en_515/adjuntos/PAPERS.pdf

PAPOYAN, Arshak and N. BAGDASARYAN (1999): *O nekotorykh voprosakh trudovoy migratsii naseleniya Armenii v Rossiyu* [About Some Questions of Labor Migration of the Population of Armenia to Russia]. Yerevan (in Russian).

POGHOSYAN, Gevork A. (1997): *Back to Homeland: Armenian Returnees from Germany*. Yerevan: IOM/ASA

POGHOSYAN, Gevork (2003a): *Migration Processes in Armenia*. Yerevan: Noyan Tapan

POGHOSYAN, Gevork (2003b): 'Theory of Modernization and post-Soviet Society: The Faith of Social Transformations in Armenia'. *Society and Economy*, No. 6, pp. 166-193

POGOSYAN, Gevork (2005a): *Sovremennoe armyanskoe obščestvo: Osobennosti transformatsii* [The Modern Armenian Society: The Peculiarities of Transformation] Moscow: Academia (in Russian)

POGHOSYAN, Gevork (2005b): *Trafficking and Labor Exploitation of Armenian Migrants: A Sociological Survey*. Yerevan: ASA; OSCE; US Embassy in Armenia

POĠOSYAN, Gevorg (2006a): *Hay hasarakowt'yownę XXI dari skzbowm* [The Armenian Society at the Threshold of the 21st century]. Yerevan: Hayastani Hanrapetowt'yan Gitowt'ownneri Azgayin Akademiayi (in Armenian)

POGOSYAN, Gevork A. (2006b): *Reintegratsiya: Sotsiologičeskoe issledovanie* [Reintegration: Sociological Study]. Yerevan: ASA (in Russian and English).

POGHOSYAN, Gevorg (2009a): 'Out-Migration from Armenia after 1990'. In: Savvidis, Tessa. Ed. *International Migration: Local Conditions and Effects*. Berlin: Osteuropa-Institut, No. 3, pp. 61-80

POGHOSYAN, Gevork (2009b): 'The Migration Flows in Armenia'. In: Düvell, Franck and Irina Molodikova (Ed.s): *Tranzitnaya migratsiya i tranzitnye strany: teoriya, praktika i politika regulirovaniya* [Transit Migration and Transit Countries: Theory, Practice, and Regulation Politics]. Moscow: University Book, pp. 193-212

Policy Brief of European Union's "Support to Migration Policy Development and Relevant Capacity Building in Armenia" Programme (2007) Yerevan: International Center for Human Development

ROBERTS, Bryan W. (2004): *Remittances in Armenia: Size, Impact and Measures to Enhance Their Contribution to Development*. Yerevan: USAID http://pdf.usaid.gov/pdf_docs/PNADB948.pdf.

SANOYAN, Dmitri and Alek D. EPSTEIN (2010): *Armenians and Jews in the 20th and at the Beginning of the 21st Centuries: Parallels of Political Fortunes*. Yerevan: Noravank Foundation

SAVVIDIS, Tessa (2009a): 'Regional Migration to Russia: Development, Interests, and Perspectives'. In: Poghosyan, Gevork. Ed. *Armenian Society in Transition*. Yerevan: ASA, pp. 149-164

SAVVIDIS, Tessa (2009b): 'Comparing out-migration from Armenia and Georgia to Moscow: Challenges and prospects of research'. In: Savvidis, Tessa (Ed.) *International Migration: Local conditions and Effects*. Berlin: Arbeitspapiere des Osteuropa-Instituts, No. 3, pp. 22-60

Sotsial'no-ėkonomičeskoe položenie Respubliki Armenii v Yanvare 2010 goda; informatsionniy mesyačniy doklad [Socio-economic Situation of the Republic of Armenian in January 2010; monthly information report] (2010) Yerevan: State Statistical Agency [in Armenian and Russian]. http://www.armstat.am/en/?nid=80&id=1125.

Statistical Yearbook 2001: Trends in Displacement, Protection and Solutions (2001) Geneva: UNHCR

Statistical Yearbook 2005: Trends in Displacement, Protection and Solutions (2007a) Geneva: UNHCR

Statistical Yearbook 2006: Trends in Displacement, Protection and Solutions (2007b) Geneva: UNHCR

TASHJIYAN, James H. (1947): *Armenians of the United States and Canada: A Brief Study*. Boston: Hairenik Press

YUDINA, Tatiana (2002): O sotsiologičeskom analize migratsionnykh protsessov [About the Sociological Analysis of Migration Processes]. *Sotsiologičeskie issledovaniya*, N 10, pp. 74-82

ZAYONCHKOVSKAYA, Zhanna A. (Ed.) (1999): *Migration in the CIS countries: 1997-1998*. Geneva: IOM

Irina Badurashvili

Out-Migration from Georgia to Moscow and Other Destinations

1 Introduction

As a typical post-Soviet country Georgia has been seriously affected by out-migration after its independence proclaimed in 1991. The 2002 population census registered a decline in population of some 20 percent in comparison with the previous census of 1989. This decrease can be partly explained by a decline of fertility but the larger part is due to emigration.

The out-migration from Georgia is mainly directed towards the Russian Federation, Turkey, United States and EU member states such as Greece, Germany, France and Spain. The Russian Federation is the major destination for Georgian migrants due to historical and economic ties as well as to geographical and cultural proximity. Turkey has been another important destination for Georgian migrants during the years following the dissolution of the Soviet Union due to facilitated travel arrangements. After the abolition of Turkey's visa requirements for Georgian citizens in 2006 and the closure of borders with the Russian Federation in 2008 Turkey has become a major destination for Georgian migrants. Moreover, there is evidence of some transit migration from Armenia, Iran, the Russian Federation and Ukraine towards Turkey and the European Union via Georgia. However, Georgia is not a key country for transit migration since its transportation system is underdeveloped and the country is not located on the most direct route between the countries of origin and the destinations of migration.

The migrant flows from Georgia towards the Russian Federation immediately following Georgia's independence were of ethnic character. They comprised mainly ethnic Russians who had previously moved to or had been born in Georgia. Gradually the share of native Georgians moving into the Russian Federation grew due to economic reasons. The increasing political tensions between the two countries and the irregular character of migration flows caused the emergence of barriers to Georgian migrants. The reasons are complex. After the dissolution of the Soviet Union the border control between the succeeding independent republics has been practically not existent for some time. This situation opened the Newly Independent States (NIS) to criminal networks and to migrants from South Asia and Africa *en route* to Western Europe. Russia and Ukraine are strategically positioned between the East and the West and have taken the burden of considerable flows of transit migrants. They either travelled on their own initiative or were trafficked in an organized manner. Following this development, in 1999 Russia requested the cooperation of Georgia to install a visa regime between both countries. In 2000 the Russian government introduced a visa regime for Georgian citizens while most of the other CIS countries continued to enjoy a visa-free movement towards the Russian Federation.

In 2006 the tensions in the foreign relations between the two countries resulted in a unilateral decision by Russia to close all her land and sea borders with Georgia. There are currently no direct transportation links between Georgia and Russia. As a response to the detention of four Russian military officers by the Georgian authorities on charges of espionage in September 2006 a number of Georgian citizens have been expelled from Russia on grounds of violating the immigration law. The 2006 deportation of Georgian nationals exemplifies the general vulnerability of migrant communities and their dependency on international relations. The recent flow of emigrants from Tbilisi, the capital of Georgia and other urban areas tends to be directed towards Western Europe and the United States. Georgian nationals continue to apply for asylum mostly in Germany, France and Austria (Migration in Georgia 2008), but applications are usually rejected.

The massive out-migration from Georgia is a recent phenomenon. During the Soviet period ethnic Georgians tended to remain in Georgia. More than 95% of them lived on the territory of Georgia. Their migration was primarily within the republic towards the capital city Tbilisi. After the collapse of the Soviet Union the citizens of independent Georgia got the chance to travel abroad without impediments. However, during the first years after the independence Georgia has been confronted with dramatic civil wars. They brought about large flows of internal displacements and inflicted social-economic hardship on the whole population. The economy was paralyzed due to lacking energy resources and the high political instability. Many Georgian citizens decided to leave the country for a better life elsewhere, emigrating in large numbers.

Similar to other post-Soviet republics, non-Georgians constituted the biggest flow of emigrants from Georgia in the first half of 1990s. The share of ethnic minorities shrank from 29.9% in 1989 to 16.2% in 2002.[1] By 2002, Greeks, Ukrainians and Jews had all but disappeared, while 80% of the ethnic Russians and more than half of the substantial ethnic Armenian population had departed from Georgia. Some Western authors (Beissinger 1996: 158) use this fact as an outstanding example of ethnic intolerance which has taken place in Georgia in that period. Specialists in Georgia point that "...on the eve of the dissolution of USSR Georgia was being led by political newcomers, inexperienced elite who tried to establish themselves at the helm by using the easiest possible way – political slogans. But some influential representatives of the political elite managed to use the slogans so that patriotism became perceived as unrestrained

1 Despite the exclusion of Abkhazia and South Ossetia from the 2002 census, these numbers seem relatively accurate (see Rowland 2006). Rowland compared 1989 and 2002 census data for the territory currently under Georgian control and found a 10% shift, from 26.3% ethnic minorities in 1989 to 16.2% in 2002.

nationalism. In those days several statements made by political figures concerning ethnic non-native population, which influenced mass consciousness, caused the feeling that a sharp rise in intolerance was happened. Uncertain and anxious, the people of different ethnic groups who did not feel themselves as "native" decided to emigrate" (Gachechiladze 1997: 27). The significant out-flow of members of non-titular ethnic groups was common for most of the newly independent states. The creation of new international borders changed the situation radically and many people belonging to minorities or suffering from economic hardship felt trapped inside the new independent countries and were eager to get out (Tishkov et al. 2005).

The economic collapse of Georgia during the first years of transition continued with side effects such as inflation, corruption, unemployment and poverty, which all contributed to the social crisis in the country. The interviewed experts suggested that the chaos Georgia endured in the early 1990s spurred particularly the emigration of highly-skilled ethnic Georgians, primarily to neighbouring Russia and Turkey. Both countries had no visa regimes with Georgia at that time (Labor Migration from Georgia 2003). Due to the prolonged social-economic crisis and the lack of prospects for improvement in the near future many Georgians continued to migrate abroad for temporary or permanent settlement. The result is the large-scale emigration flows of Georgians during the whole period after independence.

The precise assessment of out-migration from Georgia since 1990s is difficult due to the lack of reliable statistics and limited research. The official statistics on migration, particularly those before January 2004 are generally viewed with scepticism (Gachechiladze 1997; Badurashvili, Kapanadze and Cheishvili 2001; Badurashvili and Kapanadze 2003). The population census held in Georgia in 2002 allowed the State Statistical Office to reassess the inter-census population counts for the 1990s and to provide more reliable estimates of migration flows between 1989 and 2002. Some data on out-migration may be also found in the publications of independent experts (Tsuladze, Maglaperidze and Vadachkoria 2008) and in estimates of international organizations.

Since 2004 the official migration statistics is based on the data on passenger-flows provided by the Georgian Border Department. This data informs only about the gross numbers of entries and exits. There is no way to track individual comings and goings in order to distinguish migrants from other passengers. No other official statistics on migration in Georgia was published since 2003 aside of annually updated figures on net migration numbers and rates based on data from the Georgian Border Department. There is also some official data on immigration to Georgia provided by the Georgian Ministry of Justice. It is based on the information about annually issued residence permits and refugee/asylum applications' statistics collected by UNHCR.

In order to understand the peculiarity of the migration situation in Georgia and its recent modifications one must go far beyond the official statistics. This is an exercise which is interesting in scientific terms and important in practical terms. According to various estimations, the documented migration is only the top of a huge iceberg. The Georgian official statistics insists on the point that there is no evidence of large scale emigration from Georgia after 2000. The background assumption of the point is that those who wished to leave the country have already done so. The complementary assumption is that those who remained are content to remain where they are. It is further assumed that recent migration flows from Georgia are phenomena of temporary migration for seasonal employment abroad. Due to the limited employment opportunities in Georgia people go abroad to earn money and support their families at home. No doubt, the temporary moves abroad should be investigated carefully because they are numerous and of a large variety. They include diverse groups of the population and respond quickly to socio-economic and political changes. Their effect on the overall socio-economic development of the country is significant. As seen in the light of the dwindling possibilities for permanent immigration to most developed countries, these temporary moves abroad are called by some authors "incomplete migration" which is replacing traditional forms of migration. This will most likely be the dominant form of out-migration from Georgia in the near future since many developed countries increasingly need immigration.

The knowledge about migration processes in Georgia is still fragmentary. Revaz Gachechiladze (1997: 9) pointed out that "...statistics did indicate that ethnic Georgians tended to stay within the homeland, while the representatives of many other ethnic groups entered and left Georgia. According to the 1979 population census, 96.5% of ethnic Georgians in the USSR lived in the Georgian SSR, while more than a quarter of ethnic Russians and about a third of ethnic Armenians lived outside of the Russian Federation and the Armenian SSR, respectively. The abovementioned settlement pattern of ethnic Georgians is a reflection of the population's historical social structure and economic situation rather than of a specific territorial patriotism". He continued: "However, since 1957 the annual net-migration (in Georgia) turned from positive to negative. Despite this fact, net-migration figures per decade were stable and considered to be a reasonable volume. There was no total decrease in population because natural population growth more than compensated for emigration" (Gachechiladze 1997: 14). The author also mentioned that since the 1970s there was some increase in emigration flows from Georgia caused by the emigration of Jews. Despite the instructions of the all-Union authorities, this emigration was not obstructed by the Georgian government. Moreover, there was "... the understandable desire of ethnic non-Georgians to acquire further education in the native languages (Russian, Armenian, Azerbaijanian). This objective was more easily achieved in the neighbouring republics than in Georgia, where almost 90% of the curriculum in the universities was taught in Georgian" (Gachechiladze

1997:16). The tendency to leave Georgia to pursue a higher education degree in Russia was also strong among those ethnic Georgians who have attended Russian schools. Both Georgians and proportionally more non-ethnic Georgians tended to settle outside of Georgia after completing their education. This could be considered as a form of labor migration and even as a potential "brain drain". These processes caused the increase of the number of ethnic Georgians living outside of the Republic from 91 thousands in 1959 to 193.7 thousand by the year 1989.

Concerning the migration trends in Georgia after the independence, Gachechildaze pointed out that: "…it is a fact that the processes of migration in Georgia underwent dramatic changes in the beginning of 1990s. The balance of the out-migration changed radically as did the intensity of migration, especially that of emigration. People begun to emigrate to outside of the former Soviet Union" (1997: 25). He analyzed the negative and positive sides of the labor migration and mentioned such negative aspects for Georgia as brain drain and decreased birth rate. At the same time he pointed out at the economic benefits of labor migration from Georgia that "…to a certain extent lightened the pressure of unemployment here, while remittances of emigrants permitted many families in Georgia to survive during the heavy economic crisis" (1997: 55). Moreover, Gachechiladze believed that "...if the rates of economic growth in Georgia would start to rise, this will promote the opportunities for employment in this country. This development itself will reverse the migration currents and win back many 'prodigal sons'" (1997: 55). These widespread optimistic expectations of mass return of emigrants did not come about since the expected high economic growth did not materialize.

Tamaz Gugushvili (1998: 120) also points out that labor migration "…has saved Georgia from starvation". But the process deserves a special attention by the Georgian government with regard to the protection of human rights of Georgian citizens abroad. The problem is related to the status of Georgian migrants abroad. While most Georgian nationals enter foreign countries legally by using tourist visas, they end up as irregular migrants since they usually overstay the visa expiration (Irregular Migration 2000). Migrants from Georgia would certainly prefer to migrate legally for working abroad. But they have limited opportunities to do this due to the lack of bilateral labor agreements between Georgia and the countries of favorable destination of migration which would make the legal employment of citizens of Georgians abroad possible (see Badurashvili 2004).

Guram Svanidze (1998) noted that the intention to emigrate increased in Georgia in the mid-1990s. The main reason for this trend could be tracked to the unemployment and low living standard at that time. His study on the emigration of ethnic minorities has shown that the emigration intention was strongest among Armenians. However, they were less inclined to migrate to Armenia, than the Azeris to Azerbaijan. For all migrants, Georgians included, the Russian

Federation used to be the main destination. In this context Mirian Tukhashvili (1998) mentioned that already in the Soviet time some male Armenians from South Georgia used to go for temporary labor migration to Russia. They were leaving Georgia each year at spring, were jobbing as contract workers in Russia (so called "shabashniki") and used to return in the autumn to their families. In one IOM publication (Labor Migration from Georgia 2003: 7) we find the explanation that the Javakheti region in South Georgia, which is densely populated by ethnic Armenians "...has a long tradition of labor migration. The Armenian population of the Javakheti region left on a large scale to work on the new building projects in the regions of 'virgin land' in the USSR. Wages were high there because of the high level of skill required and the intensity of the work performed. Their income was several times higher than that of the people in their regions of origin. Another factor which led to labor migration from the Javakheti region was the scale of the demographic growth in the south of Georgia compared to the static number of job openings in the agricultural field, which was the center of the regional economy. Furthermore, this region was part of the border-quarantine zone of the USSR, which restricted its economic activities and hindered the effective employment of labor resources". Hence, the temporary migration from Georgia in the Soviet time comprised mainly non-Georgian population and was limited in terms of numbers and settlements. After 1991, out-migration covered the whole country. The flows of labor migration have tremendously increased due to the economic crisis.

The literature on the general characteristics of migration from Georgia has become already substantial. However, very little research has been done on the status and behavior of migrants from Georgia in the countries of destination. Only a handful of empirical studies on migration and related issues have been conducted in Georgia. The most widely referenced study (Labor Migration from Georgia 2003) surveyed the family members of 600 households with at least one member working abroad. Some data has been collected as part of annual household surveys both by the Georgian Department of Statistics (since 2003) and in the Caucasian Resource Research Center. The latter contain the results of the special surveys on remittances which were financed by the European Bank for Reconstruction and Development in 2007. There are also some periodic surveys conducted by Georgian researchers as well (Chelidze 2006). All these studies rely on reporting by relatives of the migrants. This approach is pragmatic and offers a reliable picture of the impact of the emigrants' migration on the family. But it yields no reliable data about the realities in which migrants are involved while abroad, including their earnings, type of employment, etc. Studies of returnees (Badurashvili 2005) offer a more accurate picture of the realities of Georgian migrants' life abroad and in particular of their return and reintegration experiences.

As seen from this point of view, the field studies implemented in the framework of the ArGeMi project have obvious advantages. They are based on a multi-dimensional approach to out-migration from Georgia, on the one side, and focused on one typical destination of migration, Moscow, on the other. The results of the study of Georgian migrants in Moscow may be used as a valuable orientation about the profile of Georgian migrants to Russia indeed. The political tensions in the relationships between the two countries and the military conflict of August 2008 exerted a definite impact on the results of our study. They inflated the negative assessment related to migration of Georgians to Russia. The field studies include a variety of methodological approaches like secondary analysis of statistical data, in-depth interviews with experts, structured interviews with three types of migration-related samples, and structured monitoring of events occurring during the period of study. In addition, a multi-dimensional approach was designed for the study of international migration processes, such as the ways of labor recruitment, highly qualified migration ('brain drain'), aspects of money transfer, including remittances, and prospects to return.

2 Goals of the Study and Methodology

The migration research implemented in the ArGeMi project consists of three primary components: 1) an overview of migration trends in Georgia since 1990 and profile of Georgian migrants; 2) baseline assessment of patterns of the migration flows, their profile and perspectives in Georgia, and 3) an assessment of migration policy in Georgia and the environments in which it operates with special focus on Russia. The goals of the research were the following:

A) Migration trends in Georgia:

- To reconstruct the overall picture of migration trends from Georgia since 1990 by the means of:
  1. The *study* conducted by the ArGeMi team in Georgia in 2008 was based on the analysis of documents from the national and international statistics as well as other publications, which was supplemented by the exhaustive literature review of relevant Georgian, English, and Russian language sources about or related to the out-migration from Georgia since 1990;
  2. In-depth *interviews* with stakeholders and migration experts conducted in February - April of 2009.
- To develop profiles of Georgians migrants by the means of:
  1. The structured *interviews* with returnees to Georgia conducted in February - April of 2009 with the purpose to comparatively study the experience of Georgian migrants and of returnees from Moscow in particular;
  2. The comparisons with previous *data* with the purpose to identify the developments and peculiarities in the recent migration trends in Georgia.

B) Assessment of patterns of the migration flows, their profile and perspectives in Georgia by using:

- data on return and re-migration perspectives of Georgian migrants collected during *fieldworks* conducted in February - April of 2009;

- the results of structured *interviews* with potential migrants, i.e. Georgian citizens without any previous migration experience that have been also conducted in February - April of 2009;
- experts' *opinion* on Georgia's migration's perspectives and the major factors influencing migration features in Georgia in the coming years;

C) Assessment of migration policy in Georgia and the environment in which it operates with special focus on Russia based on:

- *overview* of the migration management in Georgia, on laws and regulations on migration as well as the current return and re-integration programs for Georgian migrants;
- *data* on the international framework of out-migration from Georgia with a special focus on Russia obtained by in-depth interviews with stakeholders and migration experts conducted in February - April of 2009;
- the monitoring of migration related events in Georgia during the period of February - April of 2009 that allows to establish links between the manifest migration related events and their hidden causes and reasons, or between the manifest and the latent effects of the reporting on it in Georgia. The control round of monitoring has been conducted in January-March of 2010 in order to identify changes during the analyzed period.

Hence, the study required multiple data gathering strategies which were utilized in the reports prepared during the implementation of the project. Some preliminary results of the conducted research were published (Badurashvili 2009). In the following we will summarize the major findings of the ArGeMi project in Georgia in order to contribute to the achievement of its cognitive aims: to establish precise profiles and patterns of the current cross-boundary migration flows from the South Caucasus, to enlarge and deepen the existing knowledge and clarify the relevant details of the migration process and its consequences both for sending and receiving societies.

3 Concerns about the Data

Some concerns about the data refer to the specific situation in Georgia during the period of the field studies (1 February 2009 – 30 April 2009) were caused by the following circumstances:

- The tensions in the relationships between Georgia and Russia and the painful perception in Georgia of the military conflict in August 2008 made it difficult to keep experts focused on the migration from Georgia to Russia and not to shift to political issues concerning the relationships between both countries. The interviewed experts were subjective in their assessments and tended to exaggerate negative aspects of the life of Georgians in Moscow. The undeclared "information war" waged in the national mass-media provoked one-sided interpretations as our monitoring of events has clearly shown.

- Political tensions inside Georgia and protest actions of the opposition in Tbilisi paralyzed the normal work of major governmental institutions during the spring of 2009 and constrained the efforts to interview high level officials.
- Confrontations between government and opposition in Georgia in 2009 caused the shift of the whole public and media interest towards the local political issues and shadowed the significance of issues related to the out-migration of Georgian citizens.
- The worldwide economic crisis and the expected unemployment in foreign countries have affected the migration intentions in Georgia.

4 Migration Trends in Georgia and Profiles of Georgian Migrants

According to the most recent IOM study (Migration in Georgia 2007: 11), migration flows from Georgia are mainly directed towards the Russian Federation, the United States, Greece, Germany, Turkey, Austria, and other EU member states. A significant part of the migration to the Russian Federation is irregular with estimates ranging from 200,000 to as much as 1,000,000 of both legal and undocumented migrants from Georgia. Remittances from the Russian Federation accounted for 66 per cent of all remittances sent to Georgia and amounted to 363 million US dollars in 2006 according to the National Bank of Georgia.

The most intensive migration flows from Georgia to Russia have taken place in the period of "open borders" between the NIS countries. It is practically impossible to precisely estimate the number of people from Georgia who have settled in Russia during this period. Nevertheless, an attempt to reach a tentative assessment has been done during the in-depth interviews with experts in the framework of the present project. According to the methodology of the comparative study we have chosen for interviews 20 experts who were expected to have the necessary competence in issues concerning out-migration from Georgia. The assumption was based on the information about their professional position, position in elected bodies, political activity, personal involvement in the study or management of migration issues, and publications.

Among the selected persons were:

- 7 Higher officials and decision makers responsible for the migration management in Georgia;
- 3 Independent experts having professional experience in the migration field;
- 3 Representatives of international organizations dealing with migration;
- 4 Journalists who have published articles on the migration from Georgia;

- 3 Researchers dealing with migration research in Georgia and having scientific publications on the topic;

The first group of experts included two leading specialists of the Georgian State Minister of the Diaspora. One of them was responsible for the Georgian Diaspora in Russia and the second was in charge for the study and management of migration processes. One leading specialist in the office of Georgia's Public Human Rights Defender (Ombudsman) together with the Head of Consulate Office of the Georgian Ministry of Foreign Affairs, the Head of Demography Division in the Georgian statistical office, the Deputy Head of migration service in the Georgian Register Agency and the Head of Department for information and analysis of Georgian Ministry of Internal Affairs belonged also to this group.

The independent experts have been chosen among those having experience in the field of migration due to their previous professional activities. The group included the former Minister of Foreign Affairs of Georgia during 1995-2003. Two more persons were selected because of their interest to migration related issues due to their recent political and public activities.

The interviewed representatives of international organizations in Georgia included two specialists from the Georgian office of the International Organization of Migration and a senior officer from the UNHCR-Georgia.

Two journalists were interviewed who were mainly working for foreign media sources in Georgia. One journalist belonged to the Russian information channel on the Georgian TV and still another one was the editor-in-chief of the journal *Emigrant*.

The group of scholars comprises three well-known Georgian demographers. Two of them were professors at Georgian state universities and have numerous scientific publications on issues of migration. The third one was a researcher specializing on migration studies in Georgia, who at the same time worked in the Georgian Parliament as leading specialist in the Committee for regional policy.

We did not manage to interview any other person working in the Georgian Parliament and competent in the field of cross-boundary migration. After having cast a glance at our questionnaire, one potential interviewee surprisingly evaluated the questions as politically biased and refused the interview. Such reactions might be well understood against the background of the political tensions between Georgia and Russia.

There were three foreigners among the interviewed experts. The person working for UNHCR is a citizen of Germany, one of the journalists is a citizen of Canada and the program officer from IOM had the citizenship of the Netherlands.

The institutional affiliation and citizenship notwithstanding, the migration experts shared the same observation that there exists no reliable statistics concerning the number of citizens of Georgia who left the country after 1990 on permanent basis and concerning the number of those residing now in the Russian Federation in general and in Moscow in particular. All of them referred to estimations ranging from 800,000 to 2 million persons who left Georgia since 1990s with at least 500,000 currently residing in Russia – mostly in the big cities there. The interviewed experts were also unable to refer to any precise number of migrants from Georgia residing currently in Moscow. Some experts estimated their share from half of the migrants from Georgia living currently in Russian Federation up to 70%. The experts from the scientific community tried to be more specific by using survey data which provided the evidence that in 1989 39 thousands of ethnic Georgians lived in Moscow. Another survey documented an increase of their number up to 100 thousands in 1995. Other experts referred to the Russian population census conducted in 2001. According to this source, some 200,000 Georgians lived in Moscow. On this basis the experts estimated that there should have been at least 600,000 Georgians living in the capital city of Russia at that time since most of them were not registered during the population census.

During the interview the experts were asked to estimate the tentative number of migrants from Georgia who live and work in Moscow illegally. Whatever their institutional affiliation, the experts were not able to offer any specific figures and preferred to operate with categories like "they should not be many", "they are very few now", "difficult to say precisely", "the legally living Georgians in Moscow are more than those living illegally". They substantiated this uncertainty with the consideration that after massive deportation from Russia the Georgians living there managed to legalize their status by marrying Russians or in other ways. Nevertheless, some experts provided us with figures about the number of illegal labor migrants from Georgia living in Moscow showing "several dozen thousands", "at least 100,000 persons" or "at least 60% of those living in Russia". This range of estimations is understandable given the lack of information concerning undocumented migrants from Georgia living in Moscow.

As to the directions of recent migration flows from Georgia, all experts shared the view that Russia again is among the main recipient countries for Georgian migrants. However, the experts also expressed the hypothesis that many Georgian migrants who intended to leave for Russia finally went to Ukraine due to the restrictions in obtaining Russian visas. Moreover, some experts were sure that Georgians now leave for Ukraine primarily. Significant flows of migrants from Georgia to Turkey, Greece, Germany, and Italy were also mentioned. Other destinations like the USA, Spain, Great Britain and France are allegedly less attractive. Experts were not able to tell how many citizens of Georgia left the country in 2008 and how many to Moscow in particular. But they agreed that due to travel difficulties there was a decrease of migrants from Georgia who went to Moscow in 2008 as compared to 2007.

Estimating the gender composition of migrants, experts shared the idea that women should be at least the half of the Georgian migrants. One expert referred to the Georgian population census of 2002 showing that males and females were represented in the same proportion in the migration flows from Georgia after 1990. Some experts mentioned the higher proportions of women in particular countries, for example around 80% among migrants to Greece and Italy, while in Russia the proportion of men and women was estimated to be equal. One expert referred to his own study showing that women are more represented in migration flows from urban areas than from rural: 55-60% and 35% correspondingly. In general, all experts noted the tendency of feminization of out-migration from Georgia. Aside of specific demands for female labor abroad, this phenomenon may be explained by emancipation processes in Georgia. The new economic realities push Georgian women to take more and more difficult responsibilities.

Discussing the reasons of the present out-migration from Georgia, the experts were unanimous that Georgian migrants are mainly motivated by economic reasons. After completing their education they did not manage to find a job or they were homemakers in Georgia. They find themselves driven by circumstances to leave the country in order to provide their families with a basic subsistence. However, one scholar shared the view that during previous decades the Georgians were leaving the country in order to help their families survive. Now Georgians go abroad for improving the economic situation of the family. Young people go abroad for better education opportunities as well. Many young women try to find a partner and stay abroad for ever. Experts believed that political reasons are not so significant for recent out-migration flows from Georgia. Nevertheless, almost all experts agreed that the economic improvement in Georgia depends on the stabilization of the political situation in the country and on its international relations. This opinion was supported by the argument that the military conflict with Russia in August 2008 caused a new wave of short-term out-migration flows from Georgia.

Prioritizing the reasons for emigration, the experts put the socio-economic conditions in Georgia on the first place. In fact, unemployment represents a huge problem in the country. There are some jobs available, but not everyone can have access to them. Moreover, Georgians go abroad with the desire to get a well-paid job. The people applying for asylum abroad are allegedly motivated by the same social problems. Georgian citizens just use this way to stay legally abroad and improve their life conditions.

As to the present profile of people leaving Georgia the experts typically answered that all people that manage to get jobs abroad are allowed to leave the country. As a rule, this is the most active part of the population. Experts agree that different categories of people are involved in migration processes. They are both highly and less educated in the age from 20 to 40 years. The experts believe that they usually do not work abroad according to their education and training.

Due to the difficult economic situation in the rural areas of Georgia, the experts also assume that more and more inhabitants will leave Georgia in search of jobs abroad. These migrants mainly go to Russia and Turkey. One expert called this a new trend since it was unusual for less educated people from rural areas to go abroad. The experts also mentioned that there are some intellectuals and specialists who went abroad following invitations by business companies. But this category of migrants is rather atypical for recent migration flows. At the beginning of 1990s many intellectuals were leaving for Russia for better living conditions and earnings there. This is currently no more the case. Experts believe that it is difficult to differentiate Georgian migrants by social strata. The picture is mixed since representatives of all social categories are leaving the country due to economic reasons. Experts believe that more Georgian citizens would leave the country if they could.

The survey of returnees conducted in the framework of the ArGeMi project allows to identify the most popular destinations of Georgian migrants. Data presented in the table below does not include Russia since we did not interview in this sample (N=300) those who had migrated to Russia. We separately analyzed the data on returnees from Moscow (N=200) as the most popular destination of Georgian migrants.

Table 1: Destinations of Georgian migrants other than Russia (%) (N=300)

Destinations	Male	Female
Germany	12.1	25.2
USA	9.2	11.3
Turkey	21.3	11.3
Ukraine	7.8	3.1
Greece	9.9	18.2
Other countries	39.8	30.8

There are some gender-related preferences for migration to particular countries. This explains the prevalence of women in the migration flows to Greece and Germany and the higher concentration of Georgian men among the migrants in the Ukraine and in Turkey. This difference was identified by the experts as peculiarity related to the demands of the labor markets in recipient countries. The above list of country choice (other than Russia) corresponds with the findings of other migration studies. Data presented in Table 2 specifies the demographic profile of two categories of respondents: returnees from Moscow and those, who have been to other destinations than Russia, including the so called "far abroad" (Western European countries, USA, Turkey and Greece).

Table 2: Profile of the total sample of returnees, percentages (N=500) and a comparison between the profile of returnees from Moscow (N=200) and from other destinations (N=300)

	Total sample	Moscow	Other Destina-tions
DURATION OF MIGRATION			
Long term (more than 1 year)	36.4	39.5	34.4
Medium term (6-12 months)	17.0	15.0	18.3
Short term (less than 6 months)	46.6	45.5	47.3
SEX			
Male	48.8	51.5	47.0
Female	51.2	48.5	53.0
AGE			
Under 20	2.2	1.0	3.0
20-29	29.2	17.0	37.3
30-39	28.8	25.5	31.0
40-49	17.4	27.5	10.7
50-59	15.4	20.0	12.3
60-69	5.4	6.5	4.7
70 and more	1.6	2.5	1.0
MARITAL STATUS			
Never married	31.0	26.0	34.3
Married	56.2	59.5	54.0
Widowed	6.8	6.5	7.0
Divorced	6.0	8.0	4.7
CHILDREN			
Yes	54.6	59.5	51.3
No	45.4	40.5	48.7
EDUCATION			
Secondary general or below	22.6	23.5	22.0
Secondary professional	15.6	19.5	13.0
Higher complete	61.8	57.0	65.0
ECONOMIC STATUS			
High (Personal income over 300$)	26.4	24.0	28.0
Average (Personal income from100$ to 300$)	206	22.0	19.7
Low[2] (Without personal incomes or with income less than 100$)	53.0	54.0	52.3

Over 90% of interviewed returnees are nationals of Georgia and Georgians by ethnicity, have a permanent residence in Georgia and typically a partner, who is an ethnic Georgian. Most emigration studies reveal that emigrants are people of work age. The results of our study confirmed this. Age seems to be a highly migration-selective factor for Georgian migrants whatever the country of destina-

2 The subsistence minimum for working age male in Georgia is around 75$.

tion. Three thirds of migrants are in the age between 20 and 50 years with 28.8% of migrants in their 30s and 17.6% - in their 40s. These are cohorts that on the average are able-bodied, professionally skilled and experienced. Migrants younger than 30 years make up to 40% of the returnees from the "far abroad" and 29% among the returnees from Moscow. The mean age of returnees from Moscow in our survey consists of 42 years and from other destinations 35.9 years.

Both sexes are well represented in our samples. The male migrants are slightly more than the females (51.5% against 48.5%) among the returnees from Moscow. The percentage of women among the returnees from other destinations is higher than the percentage of males (53% against 47%). If we assume that numerous female returnees from Moscow have just accompanied their husbands or were visiting them in Moscow for a short time, then the gender composition of migration flows from Georgia to different destinations will confirm the findings of other migration studies. They have already established the fact that labor migration to Russia involves mainly the male population of Georgia, while among the migrants to the "far abroad" women prevail. This difference is related to the demand for female labor in the economically developed countries as well as to the higher price and difficulties in obtaining visas to these countries. These circumstances do not allow family members to accompany migrants or to visit them abroad.

The data collected by the interviews with returnees shows that the majority of former migrants are married with children. About half of the returnees spent abroad up to six months. More than one third of them were absent in their families longer than one year. In the migration flows from Georgia the proportion of married persons among the female migrants is significantly lower than among the males: 50% against 70% in the sample of returnees from Moscow and 45% against 65% in the sample of returnees from other destinations. Unmarried persons are stronger represented in the sample of returnees from the "far abroad" (34% against 26% among the returnees from Moscow). This may be explained by the fact that in this sample the share of those who left temporarily for studies abroad is high. This is also the reason why the age composition of respondents is different in the two samples of returnees. The under-30 year olds comprise almost 40% in the sample of returnees from the "far abroad", while among the returnees from Moscow this age group makes out only 18%. According to our survey married men stay longer abroad than singles or divorced. The proportion of those who spent more than one year abroad among married returnees exceeds 40%. Being abroad migrants did not use to visit their families in Georgia. However, they regularly maintained contacts with family members, mostly per phone.

Migrants from Georgia are usually very well educated. According to our findings, around 60% of the interviewed returnees hold a university degree and 15.6% the secondary professional level of education. In the migration flows to the "far abroad" there are more graduated persons than among the migrants to Moscow. Going abroad for work requires to be well informed about foreign labor markets, to possess foreign language skills and to be flexible in terms of territorial mobility. The well-educated stratum of Georgian society meets these requirements. This confirms the point raised by the experts that the best human potential leaves Georgia for migration. According to the ArGeMi survey, 36% of the former migrants in the sample of returnees from "far abroad" and 13% among the returnees from Moscow have a good or fluent command of English and 84% (98% in the sample of the returnees from Moscow) of Russian. These people have the ability to establish contacts in foreign countries and to adapt to the new environments. The potentials of this category of population are not used properly in Georgia and this pushes them to go abroad for jobs that may even not correspond to their qualification and experience. The data show that the work the Georgian migrants perform abroad usually does not correspond to their level of education. This holds particularly true for women. Only 5% of the female returnees (contrary to 20% of the males) mentioned during the interviews that the work they have performed abroad corresponded to their education. This fact confirms the difficult status of many Georgian labor migrants abroad as well as the limited range of working places available for migrants. They are frequently confined to non-prestigious and poorly paid jobs. On the average, female migrants from Georgia have a higher educational status than male migrants. Nevertheless, they typically perform unskilled and non-prestigious work abroad.

The prospect of better earnings is the main push factor for out-migration of Georgian citizens. Up to 35% of the returnees from the "far abroad" who had a job at the time of the interview planned to go abroad for work. The data below gives the purposes of migration:

Table 3: Percentage distribution of returnees in different age groups by purpose of the most recent migration

	Education		**Work or business**		**Visits to relatives**		**Other purpose**	
	Male	Female	Male	Female	Male	Female	Male	Female
Age	**Returnees from "Far Abroad"** *(N=300)*							
under 25	34.6	65.2	38.5	8.7	7.7	17.4	19.2	8.7
25-34	21.1	38.3	64.9	38.3	3.5	18.3	10.5	5.0
35-44	3.7	28.1	81.5	37.5	3.7	28.1	11.1	6.3
45-44	8.3	5.0	75.0	50.0	4.2	35.0	12.5	10.0
55+	-	4.2	71.4	7.8	28.6	25.0	-	-
	Returnees from Moscow *(N=200)*							
under 25	45.5	38.5	18.2	7.7	27.3	46.2	9.1	7.7
25-34	19.0	7.7	52.4	30.8	28.6	38.5	0.0	23.1
35-44	3.4	10.7	82.8	39.3	13.8	42.9	0.0	7.1
45-44	7.7	4.0	73.1	52.0	19.2	40.0	0.0	4.0
55+	-	-	50.0	33.3	37.5	61.1	12.5	5.6

Data presented in Table 3 shows that foreign education is a main purpose of migration for nearly half of all respondents under 25 years of age. Young men study predominantly in Moscow while women prefer other destinations for their studies. However, even in this young age the male returnees were abroad mainly for labor. To visit relatives is another important reason for migration from Georgia to Moscow. This reason is the most important for the female returnees from Moscow in almost all ages. The patterns of migration trips of women to Moscow might be interpreted as "migration life event history". A category of Georgian women circulates between Moscow and Georgia for visiting relatives. Being young they visit their parents living and working in Moscow. When they marry and their husbands settle there with the help of the parents-in-law, the wives visit their partners, sometimes taking children with them. Starting from the age of 35 when the children do not need too much care any longer, they try to work in Moscow themselves. In our sample more than 40% of the female migrants aged 35-44 years were working while staying in Moscow. When they grow older, these Georgian women start to visit their adult children that are by now already settled in Moscow.

Looking at the migration biography of Georgian men, we see a different pattern. Men migrate mainly for work. In some age cohorts this purpose is less dominant. But labor migration is the prevalent purpose of male migration flows from Georgia to all destinations. This also applies to the migration "far abroad" of middle aged women. The data presented in the figures below identifies spheres of employment of Georgians abroad. Female migrants from Georgia are mainly engaged in caregiving services, which appear as a typical occupation for female middle-agers.

Figure 1: Spheres of employment of Georgian migrants abroad

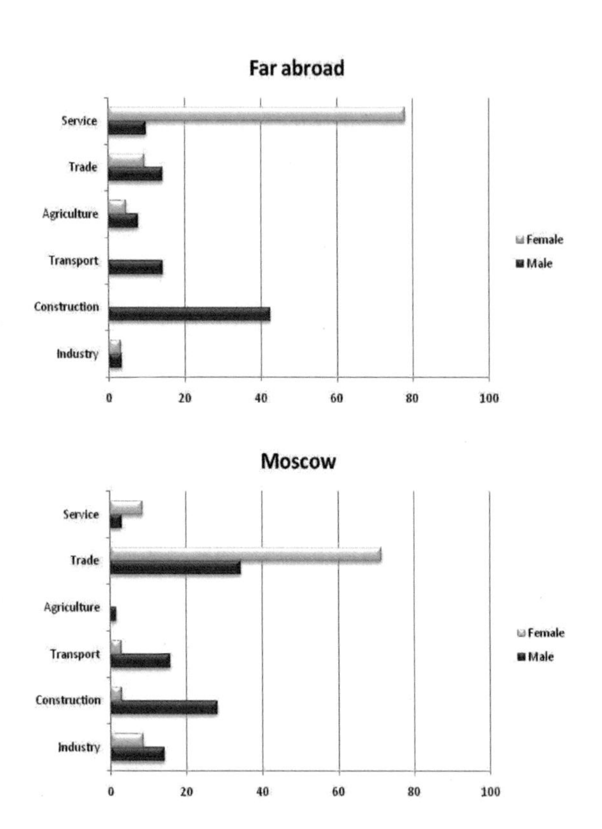

The employment of migrants from Georgia abroad has some distinctive features for those having been in Moscow and in other destination. Migrants from Georgia manage to settle better in Russia due to strong Diaspora links and longer tradition of migration to this country. The figure above shows that migrants of both sexes in Moscow are mainly engaged in trade. In the "far abroad" the labor activities are different for men and women, who are mainly employed in services, while men work predominantly in construction. But construction is a sphere of occupation that is also well represented in the sample of male returnees from Moscow, where one third of them was employed in this sphere.

The experts were well aware of the specifics of migrant labor in Moscow. They told the interviewer that Georgians are mainly engaged in trade and small business activities in Moscow. Two experts told us that this information was made public by the Moscow administration. However, one expert mentioned that the information is questionable since according to his own findings only

10% of the Georgian migrants are engaged in the Russian trade sector. Another expert mentioned that citizens of Georgia mainly live and work in big cities where criminals from Georgia settled in the 1990s and now run their own legal businesses in petrol stations, casinos, hotels, restaurants and construction. Most employees there are relatives of the owners.

The ArGeMi survey on returnees has shown that the undocumented status creates a specter of fear and anxiety. Georgian migrants in Moscow are particularly vulnerable. Over 20% of the returnees from Moscow mentioned that local authorities were unfriendly or hostile to them. One quarter of the interviewed returnees felt insecure during their recent stay in Moscow; one fifth of them were verbally offended and 8% threatened or menaced. The reasons behind such hostile attitudes the returnees explain by the fact that local people dislike labor migrants in general and Georgians in particular. Such embarrassing experience was not reported by returnees from other destinations. However, 11% of them admitted that when being abroad they felt insecure due to their illegal status and due to the fact that local people did not like labor migrants.

Nevertheless, returnees from Moscow and other destinations are generally satisfied with their recent stay abroad. The proportion of the discontent did not exceed 8% in both samples. Half of the returnees mentioned that they did not meet any hardships during their stay abroad. Those who have met some problems specified that there were difficulties mainly in getting a job and in legalizing the stay. Health problems were also noted by the respondents. It might have been difficult to deal with the health problems since only 8% of the migrants in Moscow and half of the migrants in other destinations had a health insurance during their stay abroad.

Comparing the migration experience in the two samples of returnees we would like to draw attention to the reasons behind the potential re-migration from Georgia. Twice less returnees from Moscow compared with those from other destinations plan to go abroad again in the nearest 12 months. One may assume that the returnees from Moscow have settled better in their homeland after their return and therefore abstain from re-migration. Trying to answer the question "Why returnees from Moscow are less keen to go abroad again", we compared the socio-economic characteristics of both samples of returnees and found obvious similarities in the socio-economic status of the respondents. Unemployed respondents make out up to one third in both samples. More than 70% of the households of former migrants do not receive any financial support from abroad. The economic status of all former migrants is the same. Two thirds of them have a monthly income of less than $200. The total migration experience defined by the overall number of stays abroad is also similar in both samples. Up to 60% of the respondents migrated more than once since 1990s. Hence, we do not see any other reason for diverging migration intentions in the two samples besides the more negative experience from the most recent migration to Moscow. Other studies confirm the point that the recent deterioration of the relations between Georgia and Russia aggravated the life of Georgians there. This might be a serious reason for a decision not to re-migrate. According to the survey conducted by EBRD in 2007[3] up to 40% of remittance recipients in Georgia with relatives in Russia positively answered the question "Does the conflict between Russia & Georgia affect your family members' living condition in Russia?"

At present many people in Georgia find it hard to secure a sufficient income and resort to migration to earn additional money. As a rule, economic migrants support their families and regularly send money home. When discussing the gains and losses caused by the out-migration from Georgia to Russia, all experts mentioned that the remittances of labor migrants are highly relevant. They noted that several thousands of families in Georgia have survived due to migration to Russia and the significant remittances from there. Some people started a business due to remittances. Subsequently, the impact of remittances on Georgia's economy is very beneficial. According to the experts migrant remittances contribute significantly to the country's payment balance and cover the deficit of the state budget. Money comes to Georgia and is spent in the country. Although migrants' remittances do not directly increase the state budget, they play an important role in the formation of currency reserves and help to overcome poverty. The residents of Georgia depend on remittances. Would their flow ever cease, the life in the country would become harder. Money in cash that people in Georgia receive from abroad facilitates their lives. The table below shows some differentiations of migrants by the size of the remittances they send.

Table 4: Distribution of Georgian migrants by size of their monthly remittances (%)

Amount in US$	Total		Moscow		Far abroad	
	Male	Female	Male	Female	Male	Female
Up to 100	14.4	29.5	16.3	29.6	12.9	29.4
100-200	30.6	26.9	36.7	37.0	25.8	21.6
200-300	28.8	25.6	26.5	18.5	30.6	29.4
300-500	13.5	10.3	12.2	7.4	14.5	11.8
500 and more	12.6	7.7	8.2	7.4	161	7.8

The amount of migrants' remittances varies according to the country of destination. Georgian migrants from the USA send higher amounts than migrants from Greece, Germany and Turkey. The size of average remittances of Georgian migrants from Moscow is in general lesser than from other destinations.

3 European Bank for Reconstruction and Development (EBRD) (2007): *Georgia National Public Opinion Survey on Remittances*. London: EBRD, p. 93

Figure 2: The average amount of monthly remittances of Georgian migrants

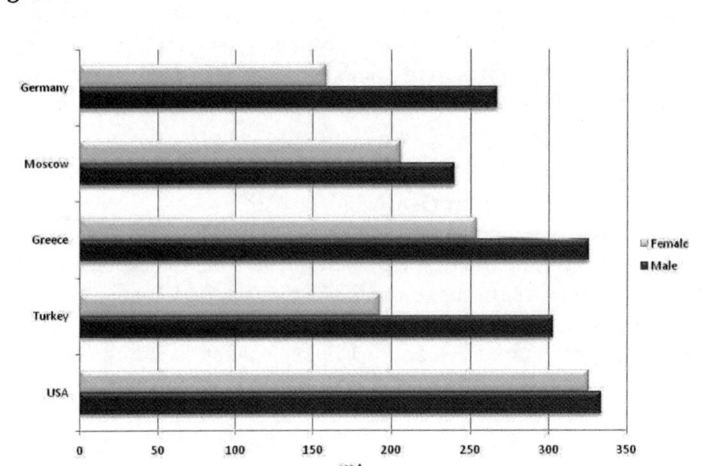

The amount of migrants' remittances is related to the incomes of migrants abroad. The living costs in the countries of destination also play a significant role in the migrants' remittances. The average amount of remittances according to our survey composed 250 US$ per month for the total sample of migrants, who were sending money to their families in Georgia. This is a sum that may provide families in Georgia only with the bare subsistence. Nevertheless such remittances allow many families in Georgia to survive against a background of massive unemployment and low local incomes.

In scholarly literature the gender differences in migrant remittances are subject of intensive discussions. In general, it is believed that women send higher portions of their income to their families than men. The ArGeMi survey has shown the opposite: In both samples of returnees the men send higher remittances than women. We can conclude that the existing gender difference in remittances of migrants might be attributed to the gap in the earnings of men and women from Georgia abroad. According to the ArGeMi study more than 70% of migrants' remittances were sent to Georgia by official bank transfers. The previous migration studies revealed the wide usage of unofficial cash money transfers by Georgian migrants. The migration survey conducted in 2005 in the framework of the international project *Enhancing Gains from International Migration in Europe and Central Asia* which was funded by the World Bank, only 38% of the migrants used official bank transfer for sending remittances to Georgia.

Survey data on remittances show that the Georgian labor migrants usually do not rely on employment mediation, but instead use informal contacts and networks abroad. More than half of the ArGeMi respondents found a job abroad through friends or relatives while 28% have arranged everything by themselves. Up to 80% of returnees found a job in Moscow with the help of acquaintances, relatives and friends. These networks replace the absence of official mechanisms for labor migration from Georgia. Over 40% of the migrants used to seek help and advice from their compatriots from Georgia. More than half of returnees from Moscow and 19% of the other returnees even stayed overnight with friends and relatives while being abroad.

5 Prospects of Out-Migration from Georgia

In the ArGeMi study we were interested in the perspectives for return migration to Georgia. Questions related to these issues were included in the experts' interviews and in the survey on returnees. Experts were asked, how many migrants from Georgia annually return and for which reasons. No expert could name any figures concerning the returnees to Georgia. The explanation is that the perception of migration in Georgia is still focused on out-migration flows from Georgia. The experts consider the scale of return flows to be insignificant. Most interviewed experts just skipped this question. One expert told that in his opinion only a small number of Georgian migrants return. Another expert mentioned that 10-15% of the migrants return in case that they could not manage to settle abroad. A third expert insisted on the point that at least two thirds of Georgian migrants return to Georgia. Ranking the typical motifs for return, the experts agreed that the reasons are mainly related to difficulties of legalizing the status of Georgian migrants abroad. In their opinion, the visa regime and the fear for illegality abroad are the most driving return reasons, aside of deportation. However, some experts believed that among the returnees there is a category of migrants who managed to make some savings abroad and think that it is a good time to return to the homeland. Other returnees are those who improved their education or qualification abroad and come back for better jobs in Georgia.

Discussing the perspectives of return, only three experts among 20 believed that most migrants from Georgia who came to Moscow after 1990 will return. These experts argued that all those who do not face legal problems in Georgia will come back. They referred to the recent trend that many former residents of Georgia who lost Georgian citizenship after obtaining the Russian one wish to re-obtain Georgian nationality. This may imply that they intend to return in future. Despite paragraph 12 of the Georgian constitution (2004) and Georgia's general unacceptability of dual citizenship there is the special Presidential order N380 issued on 9th September 2004 *On the procedure of conferring of Georgian citizenship to foreign citizens* that makes this possible. Meanwhile dual citizenship has been granted to more than 20,000 persons.

Two experts insisted on the point that the prospects for return of former Georgian residents completely depend on Georgia's economic development. Migrants from Georgia will return if they would find good conditions for working and living in their homeland. Expert N expressed the idea that it might be expected that migrants will return from Russia to Georgia, but then will leave for other countries: "By my understanding the behavior of Russia toward Georgia was awful. After all these economic sanctions, after this vandalism with shops broken, restaurants broken and after such kind of provocations, it is more difficult to remain in Russia, and after the August events it is even more difficult I think to live in a community that was always aggressive to Georgians. So, this may probably influence the return. Indeed, they would return, but to permanently stay in Georgia would be difficult. May be they will return and then go to some other places." All other experts assumed that one could not expect that many people will comc back from Moscow. They rather assumed that the majority of the migrants from Georgia may not stay in Moscow, but will stay in Russia. For those who left quite some time ago it would be difficult to come back to Georgia and adapt to the local circumstances. Moreover, many people of Georgian origin fear to come back to Georgia. "Despite the fact that they don't feel very well in Moscow, they nevertheless work and earn money there. What would happen if they lose everything with the return to Georgia?" (Expert R). "It is not likely that people, who have settled abroad and have there their own business and financial capital will decide to return to Georgia. By this decision they may lose their property and business abroad" (Expert T). Expert E mentioned that people who left Georgia at the beginning of 1990s are well settled in Russia, and they think that they may come back to Georgia after retirement. At the same time their children and grandchildren are already integrated into the Russian culture and life style. They like to come to Georgia for holidays, but not forever. According to expert D, keeping in mind that the majority of people who left Georgia after 1990 was representatives of ethnic minorities, it does not seem probable that these people would come back to Georgia. They are already well settled in Russia thanks to their previous contacts and networks.

Expert C noticed that the "Georgian government expected 150 thousands returnees to Georgia, but this prognosis was done at a time when the Georgian economy experienced an annual 10% growth. Now we have just 2 %. People may be waiting for future options. May be due to the economic crisis in Russia some will come back to Georgia, but for a short stay and then will leave again".

All experts agreed that there were many cases when former citizens of Georgia returned with the intention to stay, but then went abroad again because of disappointments. "They came to Georgia but then realized that they were not able to live in the reality they met and then they went back abroad. The majority of them has already good contacts abroad, links and networks and it was very easy for them to re-emigrate" (Expert A). Expert E told us that "...such cases had taken place in 2004 when Georgians returned after the Rose revolution, but were disappointed by the local political and socio-economic situation and went back abroad. In fact, these people returned and then left for migration again due to the same reasons as before. The reasons are again the unemployment and the economic hardships in Georgia." According to the data provided by the expert from IOM at least 20-25% of the returnees to Georgia by re-emigration programs leave again.

The ArGeMi survey on returnees revealed that the return of migrants to Georgia is mainly caused by personal reasons. This phenomenon is also known from other migration studies.[4] The personal reasons are not necessarily related to specific circumstances in the families in Georgia. They may be also linked with personal feelings and nostalgia of Georgians abroad. Migrants are often not adapted to the foreign way of life and maintain close contacts mainly with compatriots. Being unable to commute between their homeland and the country of destination due to their illegal status abroad, they miss the social environment of the home country and feel lonely. According to the ArGeMi survey, over 70% of Georgian migrants do not read Georgian newspapers and 60% do not watch Georgian TV. They do not attend Georgian clubs and maintain almost no contact with any formal Georgian organizations abroad. At the same time, according to the ArGeMi data only one third of the interviewed returnees had close friends among the local citizens abroad. Consequently, when replying to the question "If there would be free entry, which country would be the most desirable place for you to work there" the majority of the interviewees (66.5% of the returnees from Moscow and 51.7% from other destinations) mentioned that they did not have the intention to live abroad. Among those who wish to go abroad, only 12.5% among returnees from Moscow and 0.7% from other destinations mentioned Russia as desired migration destination.

The survey on 100 potential migrants conducted in the ArGeMi project demonstrated that EU countries and USA are overwhelmingly preferred, with preference given to Italy, followed by Germany, Greece and the USA. In its most recent study (Migration in Georgia 2008: 12) IOM also found that EU countries are increasingly attractive destinations for labor migrants from Georgia. The potential migrants mentioned the hope of support by friends and relatives in the preferred countries among the reasons for their destination choice. However, according to the ArGeMi survey on returnees, many employment opportunities and the good command of the Russian language appeared to be the decisive pull factors for those who migrate to Moscow. For those returning from other countries and supposedly mainly from Turkey (a neighbor of Georgia with no visa required from Georgian nationals) the bearable travel expenses and the higher wages are major pull factors.

4 Migration survey conducted in Georgia in 2005 in the framework of the international project "Enhancing Gains from International Migration in Europe and Central Asia" by the World Bank (Washington office)

Figure 3: List of desirable destinations of potential migrants from Georgia

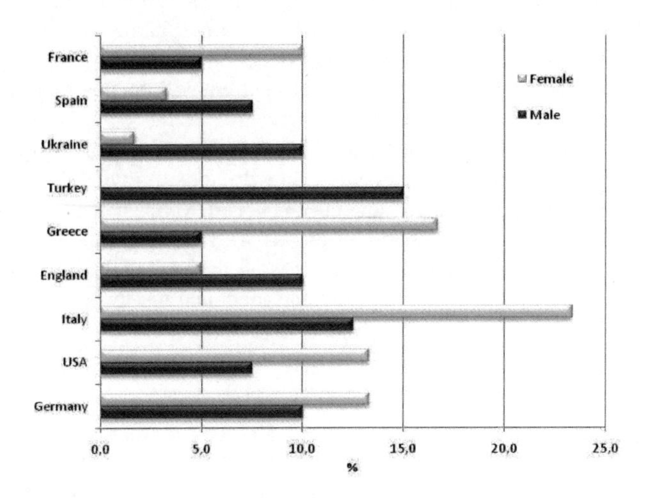

The survey on potential migrants indicates that future out-migration flows from Georgia will mostly involve young childless singles. The married persons are less than one third of the sample, but more than half of the interviewed returnees were married, with one third of young age. This shows the high migration potential among young people in Georgia who link their future career and welfare with migration abroad. Keeping in mind that more than half (53%) of the potential migrants have a higher education and 44% a secondary basic or specialized education, the future migration flows from Georgia would involve even more educationally advanced strata of Georgian society than at present. The majority of the interviewed potential migrants from Georgia (60%) speak Russian; up to 30% have command of English and another 13% of German language. The gender composition of future migration flows seems to be predominantly female, for women currently comprise 60% in the whole sample of potential migrants. Given the fact that numerous migrants from Georgia work abroad below their education and professional levels, the ArGeMi survey on potential migrants clearly indicates future brain drain combined with brain waste.

Three quarters of the respondents' households have no family member currently abroad. This means that the majority of households with potential migrants have no previous migration experience. Thus, aside of the households already practicing circular labor migration (some 10% of all households in Georgia), other households will be soon involved in migration processes. The reasons behind this perspective may be again attributed to the current socio-economic situation in Georgia. According to the ArGeMi survey half of the potential migrants intend to go abroad for work. Education is the purpose of migration for only one fifth of the respondents. Some 17% of them plan to visit relatives abroad. The potential migrants are well informed about the specific demands for labor in foreign countries. Half of those who would go abroad for the first time mentioned that they will work in the service sector. Some 15% of the interviewed intends to be employed in construction and 13% in trade. The majority of potential migrants believe that when working abroad they will be able to send at least 200$ per month to their families in Georgia. Hence, the potential migrants fully realize that their work abroad would not correspond to their education. Nevertheless, they wish to out-migrate since they do not see any other way for improving the financial situation of their families. The ArGeMi survey revealed that most potential migrants live in households with three and more persons, among whom at least one household member is an unemployed adult. Almost half of the potential migrants have no personal income. One third of their households have a total income below 200 USD. Some 31% of the interviewed potential migrants intend to stay abroad up to 6 months, 23% for a period from 1 to 2 years and 12% longer than four years.

The majority of the ArGeMi respondents are going to arrange their employment abroad through friends and relatives and 40% plan to stay with them overnight. In case of troubles abroad 46% of the potential migrants will seek help and advice from compatriots, 29% from local people they know abroad and only 8% would rely on the help of a Georgian Embassy abroad.

As previous migration studies (see Enhancing Gains from International Migration 2005) found out, many migrants reach their destination abroad by using short-term tourist visas and then overstay their terms thus neglecting the regulations associated with the duration of stay and the employment restrictions. As a consequence, many migrants are not able to commute between Georgia and the country of migration. In the sample of potential migrants up to 40% of the respondents clearly envisaged this situation in advance and mentioned that they will not be able to often visit their families in Georgia. Nevertheless, 23% of the potential migrants hope to return to Georgia at least once per year while 14% believe that they will be able to come just once in several years. Consequently, 65% of the respondents assumed that their friends and relatives from Georgia will not be able to visit them abroad as well. When thinking about hardships abroad, the majority of the potential migrants did not expect to have serious troubles. The expected difficulties concern the search for employment, the legalization of their stay abroad and the accommodation. The potential migrants hoped that the average people abroad would meet them friendly or at least indifferently.

Over three fourths of the potential migrants believed that out-migration is a curse for Georgia because the country loses its best people. However, at the same time the majority agree that migration relieves the labor market in Georgia and brings money into the country. A lesser but still significant part of the sample believed that migrants suffer exploitation and discrimination abroad while their families suffer the effects of the migration. Nevertheless, only one third of the respondents hoped that most migrants from Georgia would return to their homeland.

Discussing the perspectives of migration from Georgia, the interviewed experts on migration did not expect any significant changes, but assessed the migration trend in Georgia as stable. In their opinion, the decrease of out-migration from Georgia is obvious and will continue. However, some experts believed that out-migration from Georgia could slightly increase because of the unstable social and political situation in Georgia and because of the global economic crisis. The possible liberalization of foreign visa regimes may attract more young people to go abroad for study. As soon as the economic situation in Georgia would improve, the out-migration flows will slow down. One expert mentioned that Georgians are basically not inclined to migrate. The current situation appears as some kind of a forced migration when people have to leave for surviving economic hardships. If they would face better conditions at home they would never go elsewhere. Expert G noted that besides local conditions in Georgia the world economic development and the migration policies in the destination countries are of crucial importance for the future out-migration from the country. He expressed the view that the European countries will be more inclined to accept migrants due to the demographic situation there. This might have consequences for the out-migration from Georgia as well.

The issues related to the remittances of migrants were always in the focus of the discussions related to migration. The experts interviewed in framework of the ArGeMi project were concerned about the usage of remittances in Georgia. According to their opinion, governmental institutions and employment agencies should make efforts for more effective use of them by avoiding situations in which significant sums of money saved by labor migrants are stored in foreign banks. However, the experts also commented that this is a common problem of countries like Georgia having numerous labor migrants abroad. The effective investment of migrants' remittances would be possible under the conditions of stable economy and appropriate business environment in Georgia. The expert E mentioned that the goal of effective usage of remittances should be a central issue in countries like Georgia. She noted that "…it is necessary to create a system of state incomes from the export of labor force from Georgia and mechanisms for motivating the migrants to invest remittances in the local business." She quoted the example of Turkey that created a system of benefits for migrants sending money to the home country by introducing privileged exchange rates for opening accounts in local banks, advantages in taxation of amenities and goods brought by migrants to their homeland. Turkey also established an investment bank for workers abroad. The bank supports business development projects implemented by them. The employment at these companies is guaranteed for those who had decided to return to Turkey. The experience of other countries may be useful for Georgia, too. In the USA and in South America international networks for remittances were introduced. They significantly diminished the expenses for migrants and improved the environment for healthy competition at the local financial market. Countries like the Philippines, Afghanistan and Malaysia use modern money transfer systems served by telephone companies, internet banking system, etc. On the territory of the former Soviet Union some steps forward were done. The Russian Federation and Tajikistan signed a special agreement on bank transfers of migrant remittances.

Experts from the IOM mentioned that there were plenty of ideas concerning the better usage of remittances. The suggested innovation concerned the establishment of a special fund accumulating the money from migrants' families in order to support business initiatives of migrants in their countries of origin. Migrants might receive grants for small business and take low interest credits from the fund. An expert from the IOM provided the interviewer with information that the local office of the IOM in Georgia was negotiating on this issue with Georgian banks. The background idea was that the migrants would use the bank transfers for remittances by enjoying special low service fees. On the other hand, the periodical bank transfers of migrant remittances would be considered as a guarantee for giving credits to recipients of remittances. Other experts hold opposing views on the same issue. They argued that they cannot imagine how the government may influence on the use of remittances since these are private transfers and people should use them as they wish. They also commented that these amounts are too small for any serious investment. In addition, it is still difficult to use bank credits in Georgia successfully. Nevertheless, there was a general agreement that the remittances should be more effectively used. The expert G presented the problem as follows: "...with the probably decreasing trend in remittances size it will be very strange to teach people to spend some money for other purposes than living expenses". The expert N told us that there were stories about small businesses and enterprises based on migrant remittance. But he did not believe that the time for this has already come in Georgia. He stressed that living in Georgia was difficult even with his European salary. When living in Brussels he used to spend less on living expenses than when living in Tbilisi. He believed that remittances would be continuously used for covering living expenses in the first place. Only after the general improvement of the economic situation some remittances might be used for small enterprises. This would then push Georgian economy forward by generating employment and incomes.

The interviewed experts were alarmed by the loss of Georgia's best human potential. Due to brain-drain Georgia has already lost a lot of intellectuals, talented young people and her most active population. Provided there would be opportunities for employment in the country, the skilled emigrants could contribute a lot to Georgia's economic development. Experts noted that due to the out-migration the demographic potential of Georgia had weakened. Migration related social issues like family separation tend to cause serious problems in the family life of migrants and in the reproductive behavior of the family partners. The absence of parents negatively influences on the socialization of the children.

6 Migration Policies in Georgia

Some recent studies include a detailed analysis of the migration legislation in Georgia and its harmonization with European standards (Review of Migration Management in Georgia 2008; Migration in Georgia 2008). According to these studies Georgia does not have a written migration policy document. The existing legislation on migration issues consists of a number of laws, regulations and instructions stipulating the rights of Georgian nationals, foreign nationals and stateless persons and regulating the issues of entry, sojourn (residence), return, and irregular migration. The national legislation on migration relies on legal acts from other spheres of law like the Administrative or Criminal Law in order to ensure the prevention and prosecution of offences and crimes in the area of migration.

Several institutions are involved in Georgia's migration management. This is due to the fact that migration (immigration, emigration and residence) can have significant effects on the country's economic life, the demographic structure, the labor market, demands on social, health and education infrastructures, the environment, security and other issues. The main institutions dealing with migration management in Georgia are the Ministry of Justice, Ministry of Foreign Affairs, Ministry of Internal Affairs, Ministry of Refugees and Accommodation and the Ministry of Labor, Health and Social Affairs.

The **President of Georgia** is authorized to take decisions on the granting, removal, and reinstatement of Georgian citizenship; granting of asylum; defining the procedures for issuance, extension and suspension of Georgian visas; defining the procedure of issuing permits for residence in Georgia; and for making decisions on declaring an alien *persona non grata*.

The **Ministry of Refugees and Accommodation** is responsible for issues related to the regulation of the immigration as well as to the integration of migrants. The Ministry is charged with developing and implementing a migration management strategy and with the coordination of the state agencies dealing with migration issues. It also acts as the government focal point for the issues of internal displacement. The Ministry of Agriculture, the Ministry of Education and Science, and the Ministry of Labor, Health, and Social Affairs are also involved in assisting the IDPs. The Department for Refugees, Repatriates and Migration Issues registers asylum applications, makes decisions on granting refugee status, and is responsible for the subsequent protection and support issues. Furthermore, the Department prepares proposals and draft regulations on the implementation of internal and external migration policies, monitors labor migration and oversees projects and activities connected to the implementation of the European Neighborhood Policy Action Plan in migration matters.

The **Ministry of Justice (Unit for Migration Issues)** is responsible for the registration of foreigners, the issuance of residence permits and identification documents, as well as for decisions on and the execution of expulsions of aliens from Georgia. The *Civil Registry Agency* is responsible for registering foreigners and their residential status as well as for procedures related to the changing of citizenship of Georgian citizens. Since 2009 this agency is also responsible for procedures related to granting Georgian citizenship.

The **Ministry of Labor, Health and Social Affairs** has competencies concerning the procedures of legal labor migration of Georgian citizens abroad. However, such procedures do not take place in Georgia yet. The Ministry also supervises the administration of the State Fund for Protection of and Assistance to Victims of Trafficking.

The **Ministry of Foreign Affairs** is in charge of international agreements relating to migration like the readmissions agreements as well as of the protection of the rights and interests of Georgian nationals abroad. To some extent the Ministry is also in charge of the relations with the Georgian Diaspora.

The **Georgian Border Police** is responsible for border management and collection of statistics related to irregular migration.

The **Ministry of Internal Affairs (Special Operations Department, Unit to Combat Trafficking of Persons and Illegal Migration)** deals with combating irregular migration as well as with the investigation and prosecution of cases of trafficking and smuggling in migrants.

The **Prosecution Service of Georgia** is the central body responsible for the prosecution of migration-related offences in Georgia. They include illegal border crossing and trafficking of persons. The Prosecution Service is also responsible for overseeing the investigation activities of the police.

Against the background of the numerous measures implemented in Georgia concerning the creation of legislation and management mechanisms related to the migration, the measures for dealing with smuggling of migrants and trafficking of human beings may be evaluated as the most effective. In 2003, the first *Action Plan to Combat Trafficking* was signed by the President of Georgia. The Plan aimed at reforming laws in order to better protect the victims of trafficking, at taking preventive measures, providing assistance to victims, and implementing regular monitoring. Unfortunately, many of the foreseen measures remained unimplemented. In 2006, Georgia adopted a *Law on the Fight against Trafficking in Persons*. A national victim referral mechanism was established together with the *Permanent Anti-Trafficking Coordination Council.* It adopted the third *National Action Plan for 2007-2008* (Trafficking in Persons Report 2007). The Plan devoted particular attention to the elaboration and implementation of assistance and reintegration programs for victims of trafficking. Attention was also given to the awareness raising efforts as well as to the training of law enforcement agencies and the judiciary (Minnesota Advocates for Human Rights 2007).

In order to assess the effectiveness of existing migration policies we were interested in the opinion of migration experts. The majority of the interviewed experts shared the opinion that the Georgian government neither encourages nor discourages the out-migration from the country. The government keeps to a liberal foreign policy and abstains from intervention into migration processes. However, the political and economic situation in Georgia encourages the out-migration. Therefore, many experts shared the point of view expressed by expert O: "The government should realize that it has to create appropriate conditions for people living and working in Georgia in order to prevent that they must seek possibilities for earnings abroad". She also mentioned that the very non-existence of governmental migration policy encourages emigration flows from the country. Other experts suggested that the government should recognize the out-migration of Georgian citizens as one of its priorities in social policy and take into account this problem in the global strategy for Georgia's economic development. Expert R noted that "…for example in the USA no other policy aside of migration policy is under way. It is understandable, because the U.S. is a recipient country for many people and has special quotas for emigrants. Georgia does not pursue a migration policy as part of any policy or as a special policy, but experts recommend that there should be a Georgian migration policy at any rate, at least as a concept paper that the country may not be even able to implement in reality."

One may react to these negative assessments of the migration policies in Georgia by arguing that they are quite far from the reality and may be explained by the wide-spread perception that Georgia is losing its population in out-migration. The Georgian media widely discuss issues related to the negative effects of migration from Georgia and criticize the indifferent approach of the government concerning the problems. Even practices of a strict control of migration movements during the Soviet times were likely to be supported by some politicians who stressed the necessity to secure the general national interests. In this context the liberal approach to out-migration is a new phenomenon. The Georgian society needs time to change its stereotypes which at present are in a paradoxical state. It is widely expected that the government has to implement an effective migration policy without restraining the free movement. Some experts were not aware of recent initiatives and regulations concerning migration or did not consider them as part of the migration policy in Georgia. Other experts, who deal more closely with these issues as working in programs of IOM and UNHCR, mentioned during the interview that the Georgian government has already expressed its wish to work more on the regulation of migration, to restrict illegal migration and to assist people in finding official jobs abroad. However, such intentions do not mean that the Georgian government encourages or discourages migration. It is the general economic situation in Georgia that encourages the out-migration flows. The experts were aware of some new legal initiatives concerning the migration policies under consideration. The Georgian Ministry of Foreign Affairs together with the Minister of reintegration is currently working on the legislation related to the legalization of labor activity of Georgian migrants abroad. The project is being implemented on the basis of the experience of EU countries. Representatives from IOM mentioned that the issues related to the so called "mobility of partnership" were under discussion (see Joint Declaration on a Mobility Partnership 2009). The EU has offered to Georgia a number of pilot projects focusing on the legalization and regularization of the migration flows from Georgia as well as on the return of migrants from Europe to Georgia.

The majority of interviewed experts were convinced that no institutional framework for migration management or any central agency responsible for migration management and policy was functioning in Georgia. The often mentioned *Division on Migration* in the Georgian Ministry of Refugees and Accommodation was regarded as incapable to effectively deal with the complex problems of out-migration. According to the experts, the absence of a migration policy was caused by the non-existence of any governmental structure responsible for it. The Government and the Parliament were urged to think improvements in this field. Expert A mentioned that migration policy in Georgia does not exist because there is no political wish for it. Another expert (H) who was a high official at the Ministry of Foreign Affairs countered that actually the migration policy in Georgia is not comprehensive due to objective reasons since the country is in the process of creating her policy in general: "We have to solve the problem of creating independent governance in the process of its establishment. The problems of the regulation of migration are common for all new European countries and they try to manage it. But one thing may be pointed out: There should be a governmental agency in Georgia which is responsible for the control of the whole migration process by starting from the crossing the Georgian border, which is a crucial matter".

The experts from international organizations did not share such critical opinion about the alleged non-existence of a Georgian migration policy. But even they admitted that Georgia has not developed yet any comprehensive migration policy. For instance, the Georgian government initiated the draft of a migration policy document that is under consideration in Brussels.[5] The IOM and UNHCR offices in Georgia had commented this draft and integrated in it a refugee component. In the opinion of these experts, there still exist objections against visa facilitation for Georgian citizens, but there is also much progress under way. This document concerns the progress in registration of citizens, in the aliens' registration system as well as in the registration of refugees and their protection. Thus Georgia has committed herself to refugee protection, and significant efforts were invested for achieving this goal. Giving the refugees tempo-

---

5 Cf. *Joint Declaration on a Mobility Partnership between the European Union and Georgia*

rary residence permission, the Georgian government allows them to participate in the income generation by working legally. Expert L mentioned that up to 1,000 refugees had gained this permission. There is no action plan yet with regard to the migration from Georgia to other countries, but the situation is improving since negotiations with the EU take place. There are also initiatives for re-admission agreements between Georgia and the EU in the framework of the *Eastern Partnership Program*. Because of these initiatives the experts assumed that the Georgian government understands the need of a comprehensive migration policy. It should include the way to deal with the issues of out-migration from Georgia as well as to improve the repatriation to Georgia.

All experts were particularly asked whether there exist any migration related agreements regulating migration between Russia and Georgia. The majority of experts denied this. One expert argued that in migration there are only rarely specific bilateral regulations to be found, and Russia and Georgia are no exception. Several experts, however, named examples of bilateral regulations like the agreement on a visa regime between Georgia and Russia and the readmission agreement that was under discussion in 2005-2006.

Assessing the recent Russian policy concerning migrants from Georgia, the experts were very explicit. They described the policy as aggressive, fascist, "migrantophobe" and "Georgianphobe", especially in Moscow. It was even suggested that the violation of human rights due to this policy should become a subject of international law. The experts noted that the Russian/Muscovite policy was changing. Until recently the category of "Caucasians" was common in dealing with people from this region. Experts repeated that numerous Georgians had resided in Russia after 1990 without meeting problems before. But the situation had changed after 2000 when a visa regime between the two countries was introduced. Discussing the reasons for the negative changes the experts mentioned Georgia's refusal to participate in Russia's campaign against the Chechen Republic. After the events of August 2008 Georgians were already perceived as enemies. The attitudes and policies toward them became hostile. Expert B mentioned that the aggression against Georgians is astonishing. During the Chechen War Georgians were persecuted in Moscow even more than Chechens themselves. Implementing these actions the Russian authorities believed that they would be prompted by the return of Georgian migrants to Georgia. But this did not happen. The visa regime only pushed Georgian citizens to find ways to legalize their status in Moscow. The experts admitted that the people who left for Russia seem to feel well there, and the migration flows to Russia did not stop despite of the recent tensions between Georgia and Russia and the politically motivated deportations of Georgians from Russia in 2006 and 2007. As expert G noted: "... it may be that on the one hand there is policy, and on the other there is practice, which is not necessarily one and the same." Experts also thought that the Georgian government should re-consider its approach because the Georgian-

Russian antagonism negatively affects the situation of Georgian migrants in Russia.

At the same time, the experts agreed that the relations between the two countries were practically frozen at the time of the interviews. All experts answered that they were not aware of any new legal initiatives concerning migration policies in the Russian Federation or in Moscow. One expert even commented that against the background of the existing relations between Russia and Georgia such initiatives would be impossible.

The experts were asked to express their opinion concerning the situation of migrants from Georgia as compared to migrants from Central Asia and from other areas of the former Soviet Union. The answers diverged widely. Some experts believed that Georgians were in a privileged condition if compared with migrants from Central Asia. According to expert D "...Georgians do not have language difficulties in Russia, they speak Russian and are well adapted to the life in Russia. Therefore they are more successful in comparison with other migrants, who mainly work on non-prestigious manual jobs". Expert M noted: "Many Georgians making a serious business in Moscow come from Abkhazia. They managed to settle up well there, because they had already close contacts with Russia in the past." One expert mentioned that after the August 2008 War there was a Putin's directive not to disturb the business of ethnic Georgians in Russia because of European pressure.

Other experts believed that the situation of Georgians in Russia was worse than that of the migrants from Central Asia or from any other area of the former Soviet Union. In their opinion Russians always had phobias toward migrants. Under the current circumstances with Georgia as an enemy, Russia made the situation worse. The pro-European orientation of the Georgians was considered as treason by Georgia. These experts believed that the aggressive attitude towards Georgians in Moscow is obvious for everyone. In general, Muscovites are aggressive toward all Caucasians. These experts said that some time before the Chechens would be on top of the list of unwanted neighbors, but now Georgians are leading. Nevertheless, the experts did not agree with the assumption that migrants from Georgia were specially treated by employers in Moscow. They had not heard about such incidents and thought that migrants from Georgia were accepted by their employers in the same way as the other employees.

However, the overall majority of interviewed experts believed that Georgians in Moscow were discriminated both by the City and the Federal administration. Expert A mentioned that Georgians were treated as enemies in a direct continuation of Putin's policy towards Georgia. But expert T added that this policy does not concern Georgians only. The attitude towards members of Caucasian ethnic groups is negative in general. This hostile attitude was not determined by the recent events in Abkhazia and Ossetia alone. The major reason was the Russian campaign against the Chechen Republic. The situation was more serious in 2006 and after the August War in 2008 but had calmed down ever since. The experts supposed that the negative attitude towards Georgians did not significantly reduce their number or limit their activities in Moscow. The only effect is the increase of the "price" for living in Moscow due to the bribes that Georgian migrants had to regularly pay to the local administration.

Thus, according to the experts the perception of Georgians as enemies is a matter of fact in Moscow. The Russian propaganda is conceived as a major factor for this situation. The experts referred to sociological surveys revealing that Georgia was on top of the countries regarded as hostile to Russia in public opinion. The local population in Moscow was also described as aggressive towards Georgians. The explanation was that Muscovites were under the influence of anti-Georgian propaganda by chauvinistic parties and media, although the local population would not be able to distinguish between Georgians, Armenians and migrants from Azerbaijan. Expert N told that "in general, Russians are very interested in the exotic Caucasian nation, whose people are so different from the Slavic culture. They still are happy to be in the company of Georgians with their culture and traditions. But on the other hand, and especially after Putin's reign and with the Rose revolution in Georgia they hate and love Georgians at the same time. This is a strange behavior of the Russians toward Georgians. Russians are like punishing us for leaving them. The Georgian-Russian relation is like a love-story".

When responding to the question whether the human and civil rights of migrants from Georgia are sufficiently protected in Moscow, all interviewed experts stressed the point that Georgians in Moscow are not protected at all. Otherwise the deportation from Russia in 2006 would not have taken place. After Georgia left the Commonwealth of Independent States, the situation of Georgian nationals in Russia deteriorated. Some experts concluded that the Georgian Embassy in Moscow should have provided more help to Georgian citizens living there. The embassy should have assisted those who faced the threat of deportation. In the view of the interviewed experts Georgia is a country that is not capable to sufficiently protect its citizens abroad. Therefore many Georgians in the European countries prefer to be illegal migrants than to be somehow registered as Georgians.

In order to assess the social environments in which the out-migration processes take place we monitored all related events during the period of February-April 2009. During this period 27 events were documented, related to the out-migration in Georgia. Unfortunately, the monitoring was not successful in April 2009. Starting with that month the whole attention of the mass media in Georgia was focused on the local protest actions of the opposition in the center of the capital city Tbilisi. During the period under scrutiny there were no major publications related to out-migration from Georgia except of news, reported in Georgian newspapers, TV channels and internet pages. Two major national dailies with large circulation were selected for the monitoring. The first one was the pro-governmental *24 Saati* (24 hours) and the second one the independent newspaper *Axali Taoba* (New generation). The most informative Georgian internet page www.internet.ge and three channels of Georgian TV (one state TV and two- private channels) also had been monitored during the three months.

The main interest of the Georgian mass media during the scrutinized period was focused on issues related to Georgia's visa regimes with foreign countries. This interest is understandable. ArGeMi and other migration studies show that the accessibility of destinations significantly determines the directions of migration flows from Georgia. Almost 40% of all migration related publications in Georgia were devoted to these issues. The news concerning possible changes in the inspection of Georgian migrants at the border of Turkey and Ukraine were in the focus of the Georgian media sources together with the perspective of visa facilitation with the EU. The issues concerning the relationships between Russia and Georgia were supposed to be in the center of the interests of the mass-media, but during the monitoring period there were only few references to them. The tensions in the relationships were interpreted mainly from a political point of view which did not coincide with the aims of the research project. Moreover, the events related to Georgian migrants in Russia were typically used as pretext to blame Russia and to articulate pretensions to her. For instance, the *Congress of Georgians* living in Russia met the criticism of Georgia's politicians. The Georgian government recommended staying away from the Congress, because its leader allegedly maintained a pro-Russian policy. The headlines of newspaper publications concerning the event had a political connotation (*The Spectacle of Khubulia has the goal to split the Georgians)*. The Georgian TV channel reported that the President of Georgia in his comment on the event had mentioned that according to the Russians there were one million Georgians in Russia, while they were less. Although this event had been discussed for nearly a week in the Georgian media sources, the public was not provided with any precise information. The problems facing Georgians in Russia were hardly discussed with the exception of the request to reinitiate direct flights between Tbilisi and Moscow. The suggestions made by the compatriots in Russia for improving the bilateral relations were not publicized in Georgia.

These important issues remained unknown for the Georgian public also in the context of another event directly devoted to the relationships between Georgia and Russia. This was the video bridge organized by the news agency *RIA Novosti* in Tbilisi with the title "Russia - Georgia: Do we need each other?" The event was reported in the Georgian media in the sense that independent experts from both countries had discussed the perspectives of the relations between Georgia and Russia and particularly the ways for transforming the antagonism into a constructive approach. But no details of these discussions and proposals for improvements were publicized. It is surprising to note that an event that was initiated by the civil society in Georgia and Russia in order to improve the relations between the two countries did not find any serious interest in Georgia. The video bridge was announced only once and briefly in the pro-governmental newspaper in its column *In Two Words,* but was not reported by the Georgian TV channels. The reason might be a fear of controversial discussions. These two examples of monitoring clearly show that the Georgian mass media is far from the ideal of an objective and comprehensive analysis of events. Even in the oppositional media the coverage was focused on the political environment of the problems of out-migration and the situation of migrants abroad.

Analyzing the content of the migration related events focused on Russia and its presentation in Georgian media we conclude that even oppositional newspapers are very subjective in their reporting and lack of a profound analysis or alternative views. News are usually based on a single information source and do not create a positive public opinion. Even ordinary events related to the life of Georgians in Russia are presented in the context of political gains and losses. This trend is characteristic for the oppositional media, too. In the reports about the murder of a Georgian businessman or the arrest of four Georgians in Moscow with suspect of burglary the whole focus of Georgian media was directed towards the ethnic affiliation of these people. Their criminal background as the most obvious reason of the murder was not even mentioned.

The overall impression which remained from the monitoring of migration related events is that they were reported very briefly in the Georgian media without any updating of the information and without any analytical depth. The majority of events were reported just one time. The exception was the Congress of Georgians in Moscow that was under discussion for nearly a week. Another widely debated event was the prospect to obtain Russian visa. The topic was reported twice within ten days. During the period of the monitoring there was no significant discrepancy in the reporting by the pro-governmental and the oppositional media. Both usually referred to the same information sources. The reference to the oppositional newspaper was in fact given mostly in cases when the reporting did not appear in the governmental media. Among the 27 migration related events in Georgia there were 11 events documented on the basis of the oppositional newspaper. Among them were only three events that were not presented in governmental newspaper. One should be aware of the fact that the oppositional media in Georgia are often reporting about events that had not taken place yet. This reporting in advance confuses the general public and lacks the reliability of information about events which have already taken place. Thus, the news on the forthcoming issuing of Russian visa for Georgian citizens was publicized much earlier than it was actually arranged. Consequently, the premature publication caused misunderstandings in Georgian society.

One year later a second round of monitoring of events was carried out by using the same methodology from January to March 2010. The main purpose of this second round was the registration of changes (or lack of changes) in the national economy, politics and culture and in the respective attitudes towards out-migration from Georgia in general and particularly towards immigrants from

Georgia in Moscow. During the period under scrutiny 18 migration related events were documented that was just two thirds of the migration related events registered one year earlier. The sharp decrease in the number of registered events confirms the diminishing interest of Georgian media in issues related to out-migration. When comparing the structure of events registered in the two rounds of monitoring one may conclude that the range of events was broader during the first round of monitoring. It included cultural and social news like the new opportunities for study abroad, information about the improvement of migration management structures in Georgia and about public initiatives concerning the relationships between Georgia and Russia. However, the style of reporting about migration related events in Georgian media sources had not changed. News was again presented without analysis and the reporting did not contribute to the positive public opinion concerning sensitive issues. In both rounds the reporting on the living conditions of Georgian migrants abroad was on criminal issues as the murder or arrest of Georgian citizens. The lack of professional analysis was obvious even in comprehensive articles published in the major governmental newspaper with promising titles like "Greece at the threshold of catastrophe. How does the heavy economic crisis influence the situation of Georgians being in Greece for labor?" The author of the article did not provide any information about the consequences of the economic crisis in Greece on the conditions of Georgian labor migrants. The article just repeated well known facts concerning the delay of salaries, cutting down the number of employed and catastrophic indicators of Greek economy as discussed in foreign information agencies. As to the particular ways in which this crisis influenced on the conditions and employment opportunities of Georgians in Greece the author only mentioned an interview with the Ambassador of Georgia who tried to convince the readers that the Georgian migrants did not suffer from the economic crisis.

The news related to the relations of Georgia with Russia is extremely politicized as this was the case already during the first round of the monitoring. Nevertheless, there was some change to be found since governmental media tried to cover the relations with Russia in a more constructive way than the oppositional media. For example, the opening of the border with Russia for transit cargos from Russia through Georgia to Armenia raised extremely negative comments in the oppositional media sources. They interpreted this decision as harmful and dangerous for the national interests of Georgia. The governmental media just reported the news in a neutral way and did not comment on it.

The most important event during the second round of monitoring was the news broadcasted by TV Company *Imedi* on the evening of March 13, 2010. *Imedi* aired a deliberately false report that caused a shockwave across the country. According to the false news, Russia had invaded Georgia following a "terror attack" on the president of the South Ossetian Republic, Eduard Kokoity. The report suggested that President Mikhail Saakashvili and his government had been evacuated. Several minutes later the source "reported" the death of Saakashvili and the creation of the People's government headed by one of the opposition leaders, Nino Burjanadze. The program lasted for half an hour and included reporting about aerial bombardment of the country's airports and seaports. Only at the end the journalists at *Imedi* mentioned that this was a "special report about a possible development of the events". The reaction of the general public on this news was extremely negative. In the governmental newspaper the news were published under the title "Shock imitation". This was the only case during the second round of monitoring when public criticism was first pronounced by governmental media sources. The assessment of this event by the oppositional media sources becomes evident already from the headlines: "The reportage of *Imedi* was planned by Saakashvili", "Yellow card to *Imedi*", "*Imedi* should be closed ", "Panic fear does not save us from a real danger", "Saakashvili wishes to intimidate people", etc. Special TV talk-shows were dedicated to the analysis of the event and explained that the main purpose of the false news was to alert the general public how Russia could use a moment of political confusion in Georgia in order to forcibly change the political situation in the country. But the unfavorable reaction of the general public exceeded the expectation of the authors. The focus of the public debate shifted towards criticism of the TV channel and towards requirements to punish the responsible ones in the government who gave the permission to broadcast this pseudo news. In fact, everybody in Georgia is aware that the *Imedi* TV channel is controlled by the government and that its Director was a former Secretary General of the leading political party in Georgia. The governmental media tried not to aggravate the situation and focused the discussion on the omissions of the authors of the imitated news. The President of Georgia also limited his criticism to this point.

7 Conclusions and Recommendations

The ArGeMi study largely supported the point of view that Georgia is a country marked by a massive out-migration which is mainly directed towards the Russian Federation, Turkey, the U.S. and certain member states of the European Union like Greece, Germany, France and Spain. Following Georgia's independence, the initial migration flows from Georgia towards the Russian Federation were ethnic in character and comprised mainly ethnic Russians. There is no reliable statistics concerning the number of citizens of Georgia who left the country after 1990 on a permanent basis or concerning the number of those residing in the Russian Federation in general and in Moscow in particular.

However, there is also no evidence of a massive emigration from Georgia at present. The recent out-migration flows are mainly dominated by temporary labor migration. The citizens of Georgia usually leave for temporary labor in order to provide their families with the basic subsistence. Given the decreasing possibilities for permanent emigration to the most developed countries, the temporary migration cycle replaces the traditional emigration. This pattern will most likely become the dominant form of out-migration from Georgia since many developed countries will increasingly need migrant labor.

Migrants from Georgia are typically motivated by economic reasons. All social strata participate in the migration processes. These involve women and men, highly and less educated people mainly in the age from 20 to 40 years. Due to a difficult economic situation in rural areas, more and more people from the country-side will leave Georgia for labor, mainly in Russia and Turkey. On the whole, however, Georgian migrants are predominantly urban dwellers with tertiary and secondary education and also people having some networks abroad. There are also intellectuals and specialists who are invited by foreign business companies. But this category of migrants is atypical for recent migration flows from Georgia. The majority of migrants from Georgia are nationals of the country, Georgians by ethnicity, having a permanent residence in Georgia and a partner of same ethnicity.

The ArGeMi study revealed relevant differences in employment of migrants from Georgia abroad. The migrants from Georgia manage to settle better in Russia due to stronger Diaspora links and a longer tradition of migration to the country. The emigrants from Georgia who left the county in the early 1990s have already established their own businesses there and support their relatives and friends in adapting to local social environments and to find more prestigious jobs.

Labor migrants would prefer to migrate legally. But they have rather limited or no opportunities to do so for lack of labor agreements between Georgia and other countries. The prevalence of an undocumented or irregular status abroad creates fear and the feeling of being unprotected. Georgian migrants in Moscow feel vulnerable to the local authorities, who act with impunity. However, the ArGeMi project also discovered that returnees from Moscow and other destinations are generally satisfied with their most recent stay abroad despite the recently deteriorated relations between Georgia and Russia, which make the life of Georgians in Moscow more complicated and foster an inclination not to re-migrate.

Working abroad, most migrants send remittances to their families in Georgia. The highest amounts of remittances are being sent by migrants from the USA, followed by migrants in Greece, Germany and Turkey. The average amount of remittances sent by migrants from Moscow is less than from other destinations. Some 70% of the remittances to Georgia are currently sent by using official bank transfers. Previous migration studies used to identify the unofficial cash transfers as the usual way of sending remittances.

Going abroad for work, migrants from Georgia usually do not rely on employment agencies, but on informal contacts and networks, which replace the formalized organization of labor migration. These informal networks are very useful for Georgian migrants during their entire stay abroad. In case of troubles abroad the migrants from Georgia seek help and advice from their compatriots and even use to stay overnight with friends and relatives.

The return to Georgia is mainly caused by personal circumstances in their families, but also by feelings and nostalgia.

Currently temporary migration to the USA and to EU countries like Italy, Germany and Greece is preferred over the migration to the Russian Federation. The main factor influencing out-migration from Georgia will continue to be the difficult socio-economic situation in the country.

The interviewed experts were seriously concerned about the fact that due to the out-migration Georgia has already lost numerous intellectuals and specialists as well as many talented young people who could contribute a lot to the country's social and economic development. The experts also noted that due to the out-migration the demographic potential of Georgia has weakened. There are also such painful social issues related to migration, as family, increased problems in *par*tner relationships and the reproductive behavior of migrants. The experts expressed also concerns about an ineffective use of remittances in Georgia. According to their opinion, the governmental institutions and the employment agencies should try to organize a more effective use of remittances avoiding the current situation when significant migrant savings are stored in foreign banks.

The recent Russian policy on migrants from Georgia was described by the majority of interviewed experts as aggressive, "migrantophobe" and "Georgianphobe". They demanded to make this policy a subject of international law because it violates human rights. At the same time, experts admitted that Georgian citizens do not feel bad in Russia. Therefore, the labor migration from Georgia to Russia did not cease despite of the recent antagonism between both countries and the massive deportations of Georgians from Russia in 2006 and 2007. The experts believed that Georgians were discriminated by both the Moscow City administration and the federal administration. When responding to the question whether the human and civil rights of migrants from Georgia in Moscow were sufficiently protected, all interviewed agreed that Georgian migrants were not protected at all.

The monitoring of migration related events discovered that even the opposition newspapers in Georgia were rather subjective in their repo, without providing any profound analysis or alternative views. News was usually based on single information sources and did not facilitate the positive public opinion.

Currently Georgia does not have a comprehensive document concerning migration policy. The legislation consists of a number of laws, regulations and instructions stipulating the rights of Georgian nationals, foreign nationals and stateless persons and regulating the issues of entry, sojourn (residence), return, and irregular migration. National jurisdiction relies on legal acts of the Administrative Law or Penal Code in order to ensure prevention and prosecution of offences and crimes in the sphere of trans-boundary migration.

The main institutions dealing with migration management in Georgia are the Ministry of Justice, Ministry of Foreign Affairs, Ministry of Internal Affairs, Ministry of Refugees and Accommodation and the Ministry of Labor, Health and Social Affairs. There is no coherent system for collecting and analyzing data of trans-boundary migration. The exchange of information between the institutions dealing with migration is underdeveloped. There exists no single governmental agency dealing with the management of migration. There is no clear division of tasks between the existing agencies dealing with the migration process. In addition, the migration trends are not adequately covered by the legislation. It addresses the migration process only with general provisions which do refer to numerous specific issues. Furthermore, the legal provisions need to be clearly defined with an orientation towards the legal requirements of the European Union. In this way the Georgian migration policy should be better adapted to the management of the migration flows according to national interests and in correspondence with the country's international commitments.

There is an obvious need of a "key institution" for migration policy development. This institution should be formally charged with the coordination of the administrative activities related to cross-boundary migration. The institution should be established on the basis of the understanding that migration is a much broader issue than the influx and accommodation of refugees and will increasingly include both immigration and emigration due to economic reasons. The ongoing restructuring of the *Ministry of Labor, Health and Social Affairs* notwithstanding, there is currently no state institution dealing specifically with the issues of labor migration. The *Ministry of Refugees and Accommodation* has expressed the wish to coordinate Georgia's labor migration policy, but there is no concerted policy yet in place regulating the labor migration into and out of the country. There is no work permit system regulating the inflow of migrant workers to Georgia. This is a significant gap in the migration management system of the country.

References

BADURASHVILI, Irina (1999): 'Migraciuli procesebis shescavlisas shercheviti gamokvlevis metodis gamokeneba [The Usage of Sample Surveys in the Study of Migration Processes in Georgia]'. *Economika,* No 4-5, pp. 179-187 (in Georgian)

BADURASHVILI, Irina (2001): 'Problemy neupravlyaemoy trudovoy migracii v Gruzii' [Problems of Irregular Labor Migration in Georgia]. In: *Demography of Armenia at the Threshold of the Millennium.* Yerevan: UNFPA, pp. 5-9 (in Russian)

BADURASHVILI, Irina (2004): *Determinants and Consequences of Irregular Migration in a Society under Transition: The Case of Georgia.*

http://paa2004.princeton.edu/download.asp?submissionId=41960#search=%22%22Irina%20Badurashvili%22%22

BADURASHVILI, Irina (2005): *Illegal Migrants from Georgia: Labor Market Experiences and Remittances Behaviour* http://iussp2005.princeton.edu/download.aspx?submissionId=51259.

BADURASHVILI, Irina (2009): 'Out-migration from Georgia after 1990'. In: Tessa Savvidis (Ed.): *International Migration: Local Conditions and Effects.* Berlin: Freie Universität Berlin, pp. 80-108

BADURASHVILI, Irina and GUGUSHVILI, Toma (1999): 'Vynuždennaya migratsiya v Gruzii [Forced Migration in Georgia]. In: Zayonchkovskaya, Zhanna (Ed.): *Migration Situation in the CIS Countries.* Moscow: The CIS Research Center on Forced Migration, pp. 113-117 (in Russian)

BADURASHVILI, Irina; KAPANADZE, Ekaterina and CHEISHVILI, Revaz (2001): 'Some Issues of Recent Migration Processes in Georgia'. *Central Asia and the Caucasus*, Vol. 2 (14), pp. 220-224

BADURASHVILI, Irina and KAPANADZE, Ekaterina (2003): 'Development of National System Population Statistics and Main Trends of Demographic Development in South-Caucasian Region: Armenia, Azerbaijan and Georgia'. In: Katus, Kalev and Puur, Allan (Eds.): *Unity and Diversity of Population Development: Baltic and South-Caucasian Regions.* Tallinn: Estonian Interuniversity Population Research Center, pp. 1-48

BEISSINGER, Mark R. (1996): 'State Building in the Shadow of an Empire-State: The Soviet Legacy in Post-Soviet Politics'. In: Karen Dawisha & Bruce Parrot (Eds.): *The End of Empire? The Transformation of the USSR in Comparative Perspectives* N.Y. & London: M.E. Sharpe

CHELIDZE, Natia (2006): *Shromiti emigracia post-sabchota saqartvelodan* [Labor Emigration from Post-Soviet Georgia]. Tbilisi: Migration Study Center (in Georgian)

ERLICH, Aaron; VACHARADZE, Kristina and BABUNASHVILI, Giorgi (2009): *Voices of Migration in Georgia.* Tbilisi: The Caucasus Research Resource Center

GACHECHILADZE, Revaz (1997): *Population Migration in Georgia and Its Socio-Economic Consequences.* Tbilisi: UNDP

Georgia National Public Opinion Survey on Remittances (2007): London: European Bank for Reconstruction and Development (EBRD)

GUGUSHVILI, Toma (1998): *Saqartvelos gare migraciul-demografiuli problemebi* [The Problems of External Migration Demographic Problems of Georgia]. Tbilisi: Georgian Center of Population Research (in Georgian)

Irregular Migration and Trafficking in Migrants: The case of Georgia (2000). Tbilisi: International Organization for Migration (IOM)

Joint Declaration on a Mobility Partnership between the European Union and Georgia (2009). Brussels, 30 November

Labor Migration from Georgia (2003) Tbilisi: IOM

Review of Migration Management in Georgia: Assessment Mission Report (2008). Tbilisi: IOM

MANSOOR, Ali and QUILLIN, Bryce (Eds.) (2007): *Eastern Europe and the Former Soviet Union.* Washington DC: The International Bank for Reconstruction and Development; The World Bank

Migration in Georgia: A Country Profile (2008) Tbilisi: IOM

Minnesota Advocates for Human Rights (2007): *Measures and Actions Taken by Georgia against Trafficking in Persons* – 2007. http://www.stopvaw.org

Review of Migration Management in Georgia: Assessment Mission Report (2008) Tbilisi: IOM

ROWLAND, Richard (2006): 'National and Regional Population Trends in Georgia, 1989-2002: Results from the 2002 Census'. *Eurasian Geography and Economics*, Vol. 47, No. 2, pp. 221-242

SHINJIASHVILI, Tamar (2008): *Student Potential Migration from Georgia.* http://epc2008.princeton.edu/download.aspx?submissionId=80407

SVANIDZE, Guram (1998): 'Emigration from Georgia and Its Causes'. *Legal Journal*, No. 1, pp. 69-100

TISHKOV, Vladimir; Zhana ZAYONCHKOVSKAYA and Galina VITKOVSKAYA (2005): *Migration in the Countries of the Former Soviet Union: Policy Analysis and Research Programme of the Global Commission on International Migration.* http://www.gcim.org/attachements/RS3.pdf

Trafficking in Persons Report 2007 (2007). Washington D.C.: US Department of State

TSULADZE, G.; MAGLAPERIDZE, N.; VADACHKORIA, A. (2008): *Demographic Yearbook of Georgia,* Tbilisi: UNFPA-Georgia/Institute of Demography and Sociology

TUKHASHVILI, Miriam (1998): *Saqartvelos shromiti potenciali* [Labor Potential of Georgia]. Tbilisi: Tbilisi State University (in Georgian)

TUKHASHVILI, Miriam (2007): 'Izmeneniya v trudovoy emigratsii naseleniya Gruzii (Changes in Labor Emigration of Georgian Population)' In: *Proceedings of the International Conference Migration and Development (5th Valenteevskiye Čteniya) 13-15 September 2007*, Vol. 2. Moscow, pp. 123-135 (in Russian)

Galina I. Osadchaya

Migrants from Armenia and Georgia in Moscow

1 Introduction

Moscow of today is a huge megalopolis with a population of more than 10.5 million people. The capital city of the Russian Federation is the most developed region in the post-Soviet space and attracts mass immigration. Due to it three negative trends could be mitigated during the recent decades. The trends are the significant decline of birth-rates among the indigenous population, its fast ageing and the growing demand for labor in spheres that are not attractive for Muscovites. This demand for labor generated an annual increase of Moscow's population by 50 to 60 thousand persons (Demografičeskiy ežegodnik 2001-2009: Section 7). This rate exceeds all rates of population increase in any other constituent parts of the Russian Federation.[1] In Moscow the coefficient of the migration increment is three times higher than in the Federation as a whole. According to experts, the scale of temporary migration into the capital city in the period of seasonal peak exceeds the scale of permanent immigration by three to six times (Rybakovskiy 2006: 175). The contribution made by migrants from Armenia and Georgia to the migration increment of Moscow's population is relatively stable. Migrants from both countries come to the Moscow agglomeration more often than migrants from any other country belonging to the Commonwealth of Independent States (Zayončkovskaya and Mkrtčyan 2009).

The massive immigration increases the ethnic and cultural diversity of the capital city and thus aggravates some ethno-cultural contrasts and tensions. The deepening of the cultural gap between people arriving to Moscow and the indigenous Muscovites requires permanent scientific monitoring and well designed policies for the prevention of interethnic confrontations. This should be done in the interests of the social integration into the capital city in general and of the integration of migrants into the Moscow community in particular. For all these reasons *Reports* on the demographic issues in the capital city are being published annually, with sections on the internal and external migration. The section on the external migration provides analysis of the acceptance of migrants by the local population and of the migration policy strategies developed and applied by the Russian Federation. In addition, studies on the migration processes to Moscow are regularly conducted by migration scholars L. Rybakovskiy (demographic trends in Moscow), Ž. Zayončkovskaya and N. Mkrtčyan (migration flows

1 According to the Constitution of the Russian Federation of 1993, the Federation consists of 83 constituent parts also called subjects.

and migration policy of the Moscow City authorities), and E. Tyuryukanova (migrants in the local informal labor market). The dynamics of evaluations of the efficiency of migration policies and the changes of attitudes towards migrants were studied in the monitoring conducted by the *Russian State Social University* (Osadchaya and Yudina 2007, 2008, 2009).

The ethno-cultural peculiarities of migrants from Armenia and Georgia and their substantial contribution to the migration increment of Moscow's population often attract the attention of researchers. They analyze the push factors that drive migrants from the South Caucasus to Moscow, the employment of migrants from Armenia and Georgia on the labor market in Moscow, or the adaptation of Armenian nationals to the Moscow milieu. Furthermore, the relation between the migration of the Armenian population and the economic, political and social processes taking place in the post-Soviet space were put under scrutiny (Drobiževa, Arutyunyan and Kuznetsov 2007).

The study undertaken by the Moscow ArGeMi research team relies on a complex methodological strategy. It opens various opportunities to describe, explain and forecast the migration flows from Armenia and Georgia to Moscow. The approach also helps to develop indicators for integration of migrants into the Moscow community.

2 Aims of the Study and Its Methodology

The examination of conditions and everyday practices of migration from Armenia and Georgia to Moscow includes the analysis of the specifics of immigration to the Moscow megalopolis, the trends of migration from Armenia and Georgia to Moscow and the prospects of return migration, the integration of migrants into the Moscow community and the regulation of migration processes. The study is based on the following information sources:

a) Analysis of national statistical documents and publications of experts related to migration processes in Moscow from 1990 onwards;

b) Interviews with experts, including representatives of the Russian and Moscow authorities, the *Federal Migration Service* of Russia (FMS), of non-governmental organizations and the academic community. All in all, a total of 25 experts were interviewed in the period from 1 February 2009 to 30 April 2009. The selection of experts was done by a formalized choice and by the snowball method. The first group consisted of officials from the *Ministry of Regional Development*. They have numerous scientific publications on migration issues. The group also included officials from *Department of Organization of Employment and Labor Migration* of the Moscow City Employment Service. The second group of experts included a chief of the *FMS Directorate for Moscow City* and two leading experts from the *Department for Migration Registration* and from the *Department for migration Registration and Application of Administrative Laws* of the FMS Directorate for Moscow City. The third group of experts included the president of the Union of Migrant Communities in Russia, the chairman of the *Labor Migration* association, a lawyer from the Armenian community, the chairman of the *Public Council of the FMS* and a member of the *Scientific Council of the FMS*. A journalist writing on migration issues was also included in this group.

The scientists comprised the largest group of interviewed experts. The group included doctors of sciences, professors in charge of centers for interethnic relations studies, of demographic studies, of xenophobia studies and prevention of extremism in the *Institute of Social and Political Studies* and in the *Institute of Sociology* at the Russian Academy of Sciences. The group also included chairs of population research at the *Institute for Social and Political Studies*, the *Moscow State University*, the migration processes management chair of the *State University of Management*, heads of laboratories of population economics and demography. All these scholars have numerous publications on migration issues, including studies on the immigration from Armenia and Georgia.

Most interviewed experts were RF nationals, although they are not necessarily ethnic Russians. Scholars of ethnic Armenian and Georgian origin took part in the interviews. Depending on the area of their qualification or engagement, the interviewees explained their knowledge of migration processes in Moscow either by their own direct involvement in the study of these processes, by their professional positions or by their positions in legislative or executive bodies of power.

c) Structured interviews with migrants from Armenia and Georgia who arrived to Moscow after 1990. 200 migrants from Armenia and 200 migrants from Georgia were selected by applying the snowball method. The results of the polls cannot be projected on migrants from these two countries who came to Moscow before 1990.

d) Monitoring of events that occurred in the period from February to April in 2009 and in 2010 allowed to track the development of migration processes in Moscow and their coverage in newspaper articles, documents of governmental agencies, economic, political and cultural institutions, in publications of research centers, non-governmental organizations and international agencies. The aim of the monitoring was to identify and document events related to migration from Armenia and Georgia and to analyze the events as well as the kind of their coverage. Three major events added to the social context in which our study was conducted in 2009. First, this was the aftermath of the military conflict between Georgia and Russia in August 2008. Second, a speech by M. Saakashvili, the President of Georgia caused information clamor. He said that the main goods exported by Georgia to Russia were not wines but criminals. Third, the global financial and economical crisis became manifest in Moscow in early 2009.

3 Specifics of Migration Processes in the Moscow Megalopolis

In the past immigration was an important source of population replenishment in Moscow and it still performs this function nowadays. A particularly swift population growth started after Moscow was proclaimed capital city of the Soviet government in 1918. According to the 1897 population census Moscow had 1.04 million residents. In 1926 the population of the capital city comprised already 2.03 million and in 1939 it was 4.6 million. In 1959 the population of Moscow exceeded 6.1 million. The census of 2002 recorded 10.4 million persons. Therefore the number of Muscovite residents increased by the factor ten in the course of 105 years (Territorial'nyi organ Federal'noy služby). The Moscow agglomeration is an attractive destination for cross-boundary migrants. The analysis of the natural loss of population shows that the dynamics of the population there has been dominated by the migration generated increment. In recent years the influence of migration is increasing. While in the 1960s migration provided for half of the capital population growth, from 1989 onwards the population increased exclusively due to migration.

The dissolution of the USSR abruptly aggravated the international, interethnic and confessional relations in the territory of the former Union. The dissolution also worsened the economic, political and social conditions in most of the newly established sovereign states. As a result, principal changes in migration processes in Moscow came about. The social-demographic and social-ethnic characteristics of the migrant flows became rather different. The predominance of internal migration processes due to educational, marital and labor reasons in the 1980s gave way to migration flows from the new independent states during the 1990s. Forced migration and flight became part of the new phenomena.

According to statistical data, the migration balance was negative in Moscow in the early 1990s (-17,118 persons in 1991-1992). Several interviewed experts assumed that this trend might have been connected with the privatization of housing. Some Muscovites sold their apartments and moved to other cities or to the countryside. However, other experts did not agree that the official data reflected reality. They believed that the registered trend was an artifact caused by changes in the procedure of population registration (for instance, servicemen were excluded from the migrant registration in 1992). Still other experts related the trend to the decline of marital and educational migration after the profound political changes in the Russian Federation. Finally, some specialists explained the trend by the harsh measures taken be the Government of Moscow in order to prevent the massive influx of forced migrants to the capital city (Čislennost' i migratsiya naseleniya 1998: 103).

The year 1994 witnessed a surge of migration and the beginning of net balance growth in the mechanical increase of Moscow's population. The migration influx from the former Soviet republics to the RF capital city increased. Most of the migrants were ethnic Russians who moved out from the new sovereign states due to the instability of the social and economic situation, widespread nationalism, and intolerant attitudes towards persons who did not belong to the titular nation. As a result, in 1998 there were 17.2 migrants per 10 thousand indigenous residents of Moscow. Nevertheless, this figure was still 4.4 times less than the average figure for Russia as a whole. The trend of massive migration inflows to Moscow continued thereafter (Demografičeskiy eżegodnik 2001-2009: Section 7):

Table 1: Migration generated increment of Moscow population in 2000-2009 (1000)

Year	The total migration increment/loss	Migration increment from the CIS and Baltic States
2000	65,5	16,6
2001	52	6,8
2002	48	6,1
2003	54,1	3,3
2004	51,7	3,599
2005	54,9	7,608
2006	50,7	9,107
2007	50,9	9,051
2008	55,1	10,337
2009 (9 months)	38,14	-

From 2001 onwards a tightened migration policy brought about the decline of the registered migration increment flow and simultaneously the growth of irregular migration. Only migrants who were naturalized as citizens of the Russian Federation prior to their arrival in the capital city could legally settle in Moscow. Hence, the census of 2002 revealed the underestimation of the size of the migrant population. According to experts, the underestimation comprised as many as 1.5 million persons, who came to Moscow during the 1989-2002 period. The results of the census were called into question. In subsequent years, a number of studies were carried out which confirmed that a huge migrant population was permanently present in Moscow. Most migrant residents did not have the chance to legalize their stay due to the existing migration laws and also due to the corruption of officials (Smidovich 2005: 39; Kozakov 2005). Many of these problems still remain unresolved. It will not come as a surprise if the census of October 2010[2] would be overshadowed by them.

The migration influx changed the ethnic composition of Moscow's population. Between 1989 and 2002 the share of ethnic Russians in the metropolitan population declined by 5%. This change signaled the increase of the ethnic diversity. It mostly came about due to the inflow of migrants from the South Caucasus and from Central Asia. The number of ethnic Armenians and Georgians in Moscow increased by a factor of three. The number of Azerbaijanis has even

2 First results will be available only at the end of 2011.

increased by a factor of 4.5 during the period under scrutiny. According to the 2002 census the combined share of Azerbaijanis, Armenians and other 'Caucasians' in the population of the capital city already comprised 3% (Drobiževa, Arutyunyan and Kuznetsov 2007):

Table 2: Ethnic structure of the population of Moscow (2002 census)

Self-identification	Number (persons)	Percentage
Russians	8,908,854	84.8
Ukrainians	252,113	2.4
Tatars	168,075	1.6
Armenians	126,057	1.2
Azerbaijanis	94,542	0.9
Jews	84,038	0.8
Byelorussians	63,028	0.6
Georgians	52,524	0.5
Persons who did not indicate their ethnic origins	420,189	4%

In the past 10 years the statistics recorded some fluctuation of the migration increment of Moscow's population at a level of roughly 50 thousand per year. An increase in international migration was registered since 2004 onwards. The experts used to explain the decline of the increment observed in 2001-2004 by legislative changes. However, they were unable to offer any comprehensive explanation for the migration increase in 2004-2007. It is usually emphasized that the liberalization of the migration legislation since 2007 included the statistical registration of foreign citizens and stateless persons who for the first time in RF legislation got permissions for a temporary stay (Demografičeskaya situatsiya 2009). As a result of this legal innovation the number of immigrants who arrived in Moscow in 2008 increased by 6,817 persons in comparison to 2006 (Demografičeskiy ežegodnik 2009: Section 7):

Table 3: The total migration to Moscow in 2006-2008 (1,000)

Year	Total number of new residents	Of these new residents from abroad	Total migration increment/loss	Migration increment due to exchange with the CIS and Baltic countries
2006	81,131	11,879	50,600	9,107
2007	81,022	11,865	50,800	9,051
2008	87,948	13,256	55,060	10,337

The statistics shows that the lesser the influx of migrants to Russia is the greater is the share of migrants who concentrate in Moscow (Demografičeskiy ežegodnik 2009: Section 7):

Table 4: Coefficients of migration generated increment by constituent parts of the RF (annual migration increment per 10 thousand of population)

	1990	1995	2000	2003	2004	2005	2006	2007	2008
Russian Federation	19	44	25	6	7	9	11	18	18
Central Federal circuit	55	11	82	36	34	35	39	46	50
Moscow City	176	260	239	52	52	53	48	49	52

The current migration influx to Moscow covers virtually all kinds and purposes of migration like resettlement, seasonal, push-pull, occasional migrations and also all categories of migrants like economic, forced, familial, educational etc. The migrants from the three South Caucasian republics proved to be the most competitive. According to data provided by Ž. Zayončkovskaya and N. Mkrtčyan, the share of new arrivals from these countries among migrants who settled down in Moscow is 1.5 times higher than among migrants who settled down in Russia or in the Moscow region (Zayončkovskaya and Mkrtčyan 2009):

Figure 1: Specific weight of immigrants who came to Moscow from the former Soviet Republics in 2008 (in %)

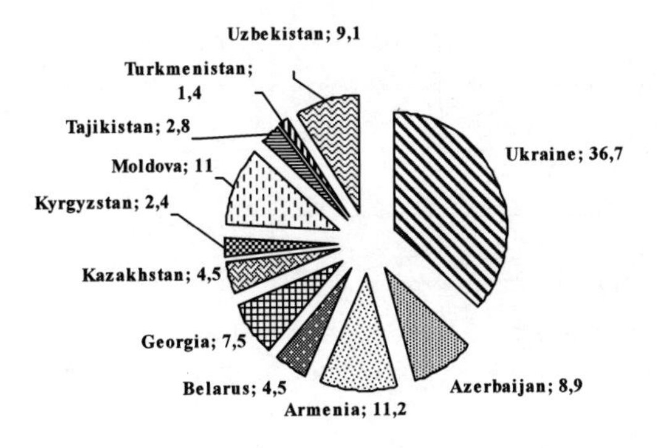

The changes in the ethnic composition of migrants brought about rejuvenation of migration flows and deepened the ethno-cultural gap between the migrants and the indigenous residents. As a rule, migrants from Central Asia come from rural areas. They have no professional qualification and only a poor command of Russian. They are frequently subject to exploitation and often become targets of humiliation and attacks.

The legal employment of migrants in Moscow continues to be an issue. According to various estimates, up to 1.5 million citizens of the former Soviet republics came legally to Russia and stayed in Moscow. Most of these newcomers work without permits as required by law (Moskovskiy meždunarodnyi portal 2009). According to expert estimates, in 2007 the legally employed foreign citizens comprised about 8% of the employed population in Moscow. In the construction industry their share was 19% (Tyuryukanova 2009). The experts estimate the share of irregular or illegal migrants is about 12 or 14% of the overall migrant labor force. The age of migrants ranges up to persons in their 50s. Elderly migrants have specific difficulties in adapting to the labor market in Moscow. At present men predominate in the gender composition of migrants in Moscow.

Another specific feature of the immigration to Moscow is the fact that migrants who work in the capital city are increasingly settling down in the territory of Larger Moscow[3], whose population comprises up to 15 million people (Polian and Selivanova 2005: 290). The reason lies in the high expenses of housing in the capital city (Demografičeskiy ežegodnik 2009: Section 7):

Table 5: Migration generated increment, 2008 (persons)

Area	**Overall migration increment**	**In result of**			
		Internal RF migration	Including		Migration exchange with foreign countries
			Movements within Regions	Movements among regions	
Russian Federation	242,106	-	-	-	242,106
Moscow Region	75,267	55,316	-	55,316	19,951
Moscow City	55,060	44,723	-	44,723	10,337

The migration processes to Moscow are influenced by the international relations with the migrant sending countries. There are good relations between Russia and Tajikistan or between Russia and Armenia. In recent years the relations between Georgia and Russia were marked by conflicts. The Russian-Georgian confrontation in 2006 and the military conflict between both countries in August 2008 had a negative impact on the public perception of migrants from Georgia. After the 2008 military confrontation, half of the RF respondents have changed their attitude towards migrants from Georgia for the worse (Vserossiyskiy tsentr 2008). Georgia's formal withdrawal from the CIS has certainly complicated the search for coordinated solutions of the migration issues and will have a negative impact on the legalization of migrants from Georgia.

3 Larger Moscow consists of the Moscow region.

The Federal Law 110 *On the registration of foreign citizens and stateless persons in the Russian Federation* (issued on 18 July 2006) introduced a notification system in the registration of foreigners at places of their residence. The Law also changed the procedure for legalizing the employment of foreign citizens. These legal amendments brought about a growth of regular migration and a reduction of the irregular migration. The authorities in Moscow nevertheless decided to close Moscow for those who want to legally reside there. In 2009 the procedure of issuing work permits to foreign citizens was further restricted. Moscow reduced the annual quota of employment permits to foreigners by 30%. According to Konstantin Romodanovskiy, the Director of the *Federal Migration Service*, this measure protected the local workers from the excessive competition of foreigners. In a meeting with Prime Minister Vladimir Putin he further mentioned that 34 thousand of foreign nationals were deported from the Russian Federation in 2009. This meant an increase of the number of deportees by 70% as compared to their number in 2008 (Romodanovskiy 2010). In its report on 2009 the FMS emphasized that it paid much attention to the fight against the irregular migration. In reality, a large number of irregular migrants avoided registration. According to Yuri M. Luzhkov, the former Mayor of Moscow, four of five labor migrants employed in Moscow did not have a proper work permit in the beginning of 2010 (Interfax 2010). K. Romodanovskiy confirmed this by adding that in 2009 the total of legally working migrants was rather insignificant.

The Moscow migration service registered 1,787,796 persons as migrants in the capital city in 2009. In the period between January and October 2009 244 thousand foreign nationals got work permits in Moscow. This was nearly twice less than in the same period of the previous year when the migration service issued 433 thousand of work permits. The greatest amount of these registered migrants (65 thousand) was employed in trade, in the construction industry (37 thousand) and in transportation (17 thousand). According to data provided by the *Department of Labor and Employment*, only one third of migrants in the capital had applied for work permits. No less than 100 thousand foreign workers were employed on construction sites in Moscow (Krizisnaya migratsiya…2010). The conclusion might be that at least 63,000 migrants were employed without a work permit in this sector of Moscow's economy alone. This makes obvious the inefficiency of the quota regulations for foreign workers. In 2009 it was cut down to 200 thousands. Transportation firms were forbidden to hire immigrants from the CIS countries as drivers. However, some professions and skills will be exempted from quotas. Specialists having these professions and skills will be allowed to get work permits. The experts are unanimous on the point that the reduction of work quotas practically increases the shadow employment. Under the circumstances of economic instability, the restrictive quota policy also increases the need for short-term labor migration. It will most probably impede the substitution of the loss of able-bodied population due to the low fertility of the local population.

4 Migration from Armenia and Georgia to Moscow - Trends and Prospects
The disastrous earthquake that hit one third of Armenia in December 1988 started the massive migration from Armenia to the Russian Federation in general and to Moscow in particular. The collapse of the USSR strengthened the migration flows. The highest influx of migrants from Armenia to Moscow was observed during the military conflict in Karabakh (1988-1994). The greatest influx of migrants from Georgia coincided with the ethnic conflicts in South Ossetia in 1991 and in Abkhazia between 1992 and 1994 (Mukomel 1999). From 2000 to 2004 the numbers of registered migrants arriving from Armenia and Georgia to Moscow constantly decreased. In 2000 2,567 persons from Armenia and 3,030 persons from Georgia were registered in Moscow. In 2004 they were only 357 registered new arrivals from Armenia and 541 from Georgia (Demografičeskiy ežegodnik 2010: Section 7). The decline of the migration generated increment of population due to migrants from Armenia and Georgia in this period may be explained by return immigration of fellow nationals. In 2002 new laws (*On the citizenship of the Russian Federation; On the legal status of foreigners in the Russian Federation*) were issued that hampered the procedure of immigrant registration, changed the rules of statistical accounting and caused a reduction of the registered migration.

A new wave of migration started after 2004 and the migration generated increment of the capital population began to rise. Migrants from Armenia and Georgia made their contribution to this trend. In 2004-2008 the increment of the migration from Armenia increased by the factor of 4.8. The migration increment from Georgia rose by the factor of 1.9. In 2008, the migration generated increment of population in the migration exchange comprised 1.2 thousand persons from Armenia and 0.73 thousand persons from Georgia. The growth of the absolute numbers of arrivals from Armenia and Georgia and their share in the total number of arrivals to Moscow from foreign countries in 2008 bear witness to increase of immigration in comparison to 2006 (Demografičeskiy ežegodnik 2010: Section 7):

Table 6: The total net results of migration to Moscow in 2006-2009 (persons arrived)

Year	The total number of arrivals to Moscow	Of these from Armenia		Of these from Georgia		Arrivals to Moscow from foreign countries	Of these from Armenia		Of these from Georgia	
	Total	total	%	Total	%	Total %	total	%	Total	%
2006	81,131	1,020	1.26	705	0.9	11, 879	1,020	8.6	705	3.8
2007	81,022	1,172	1.45	789	1.0	11, 865	1,172	9.9	789	6.6
2008	87,948	1,348	1.5	905	1.0	13, 256	1,348	10.2	905	6.8
2009, 9 months	60,738	1,034	1.7	654	1.1	9, 266	1,034	11.2	654	7.1

In 2009 the migration increment due to exchange with Armenia increased, while the migration increment obtained due to the exchange with Georgia remained at the level of 2008. This fact can be explained as consequence of the military confrontation between Georgia and the Russian Federation in August of 2008 and by the imposition of a visa regime by Russia. According to data of the *Federal Migration Service*, the number of migrants who received a status of forced migrants from Georgia comprised 19% of the total number of such migrants in 2009.

The interviewed experts admitted the incompleteness of all above mentioned data concerning the registration of migrants. According to their opinion, it is impossible to achieve a precise registration of the migration generated increment of Moscow's population on the basis of the available official data. The estimated numbers of persons who left Armenia and Georgia for permanent residence in Moscow after 1999 are rather divergent depending on the source of information. The estimations range from 50 thousand to 1.5 million of migrants from Armenia and from 25 thousand to one million of migrants from Georgia. An expert who was responsible for all migration issues in Russia during the 1990s commented that in this period Moscow received more than 50 thousand migrants from Armenia, several tens of thousand migrants from Abkhazia and more than 30 thousand migrants from South Ossetia. The discrepancies in numbers might be interpreted in the sense that the actual increment of migrants from Armenia and Georgia exceeded the registered increment by factor of 2 or perhaps more. The experts estimated the number of naturalized migrants from Armenia and Georgia also differently. For migrants from Armenia estimates ranged from 70 to 80% of persons who immigrated to Moscow on permanent basis. For the migrants from Georgia the estimates ranged from 50 to 80% of the migrants from Georgia who immigrated to Moscow.

Analyzing the immigration from Armenia and Georgia, the experts noted the shift from politically, culturally and socially motivated immigration to the predominance of economic motifs. The shift occurred gradually in the period from 1990 to 2008. In their interviews the experts compared the motifs for immigration from Armenia and from Georgia to Moscow:

- The most important motifs for immigrants from Armenia are the search for a job, the opportunity to earn more than at home, education, reunification of families, the migrant's hope that he/she will find a better business environment, and the search for a better quality of life;
- The most important motifs for migrants from Georgia is also the search for a job, followed be the flight from political instability, disagreement with the policy of the Georgian leaders, the creation of new businesses, reunification of families, and the search for a better quality of life.

These assessments were substantiated by data about the ethnic composition of the migration exchange. From 2000 onwards ethnic Armenians predominate in the structure of migrants from Armenia. The structure of ethnic migration from Georgia is different: Russians, Georgians, and Armenians are represented in nearly equal proportions. In 2006-2007 ethnic Armenians comprised the greatest share in the migration inflow from Georgia and the majority of those migrants, who arrived to Moscow and stayed there. The migrants from Georgia and Armenia, whom we interviewed, described the purposes of their arrival the search for employment, education, and permanent residence:

Table 7: Purpose of respondents' current stay in Moscow (multiple choices)

	Migrants from Armenia (%)	Migrants from Georgia (%)
Employment	52.3	57.5
Education	24.6	32.5
Permanent place of residence	13.5	11.6
Entrepreneurship	3.0	10.1
Visit to relatives	2.0	2.0
Refugee	1.0	1.5
Medical treatment	0.5	1.0
Other purpose	2.0	5.0

There was an amazingly wide range of experts' opinions concerning the profile of the migrants from Armenia and Georgia. Some interviewees believed that migrants from Armenia are highly educated and urban dwellers while others used to define them as mostly undereducated villagers. The most rigorous assessment was given by the chairman of the *Union of National Communities of Russia.* According to him 38-45% of migrants from Armenia were well educated, while undereducated people comprised some 45%. He also pointed out that some 60% of these migrants were urban dwellers in Armenia and some 40% were villagers. Last but not least, he estimated that some 97% of the migrants were ethnic Armenians and some 3% were representatives of ethnic minorities in the country. Most experts believed that the majority of migrants from Georgia consisted of representatives of ethnic minorities. They frequently used the term "undereducated" to characterize the estimated educational average of migrants from Georgia.

The ArGeMi field study discovered that 55.5% of the respondents from Armenia and 54% of the respondents from Georgia had the citizenship of the Russian Federation while 2.5% of the migrants from Armenia and 6.5% of the migrants from Georgia had a double citizenship. The overwhelming majority of the migrants from Armenia (99%) identified themselves with the titular nation, whereas only 39.5% of the migrants from Georgia belonged to the titular nation. This latter finding confirms both the high degree of ethnic diversity in Georgia and the high degree of ethnic non-Georgians among the migrant community from the country.

The data obtained in the course of the ArGeMi interviews lead to the conclusion that immigrants who came from Armenia and Georgia to Moscow after 1990 have a level of education which is equal to the average level of education of the Muscovites. Some 43% of the migrants from Armenia have secondary education and 42% have higher education. Some 47% of migrants from Georgia have secondary education and 34% have higher education. These figures coincide with or even surpass earlier estimates:

Table 8: Education of migrants from Armenia and Georgia in Moscow (%)

Respondents	ArGeMi study, 2009 (data of interviews)		Arutyunyan, Drobiževa and Kuznetsov 2007		Lebedev 2006	
	Higher education	Secondary education	Higher education	Secondary education	Higher education	Secondary education
Migrants from Armenia	42.0	43.0	39.9	20.8	36.0	Data Unavailable
Migrants from Georgia	34.5	47.5	37.3	22.7	32.0	Data Unavailable
Moscow (for reference purpose)	Higher education: 41.5% Secondary education: 49.2% Karpenko, Bershadskaya and Voznesenskaya 2008					

Women have a higher level of education. Some 48.3% of the female migrants from Armenia and 37.75% of the female migrants from Georgia underwent higher education.

The ArGeMi survey revealed that the overwhelming majority of migrants stem from cities and towns. Some 39% of the migrants from Armenia and 34% of those from Georgia came to Moscow from the capital. Some 20.5% and 25.0% of migrants respectively came from other big cities. Mere 3.5% of the migrants from Armenia and 4.5% of the migrants from Georgia came from villages. Thus, five of every 10 migrants from Armenia and Georgia might be basically well prepared to work and live in a big city.

The respondents were predominantly young people in the active working age. Seven of 10 migrants were younger than 35 years. This has positive impacts on the age structure of Moscow's population. The age cohort from 36 to 55 years comprised 25.5% of the migrants from Armenia and 33.5% of the migrants from Georgia. Education was the main purpose of the young migrants' move to Moscow. Eight of every ten migrants from Armenia and six of ten migrants from Georgia in the age cohort of 18 to 22 years came to Moscow for education. Referring to the information from the interviews, we may say that 55% of the migrants from Armenia and 48% of those from Georgia who arrived to Moscow after 1990 were single men. Some 36% and 38% of the migrants from Armenia and Georgia respectively were married. Tentatively every tenth was divorced. There are less unmarried and more divorced or widowed women among the migrants from Armenia and Georgia than among the male migrants from both countries. The majority of migrants from Armenia (55%) and Georgia (58.5%) did not have households or children in their country of origin. Six of ten respondents in both groups did not have to care about children. However, 35% of migrants from Armenia and 41% of migrants from Georgia had to take care for children in the big families that they had left in their countries of origin. This is a substantial economic burden since there were no employed able-bodied relatives in every tenth family of migrants from Armenia and in every fifth family of migrants from Georgia.

About half of the migrants from both countries transfer money to their families, relatives or friends in their country of origin. These remittances obviously play a key role in the efforts of the population of Armenia and Georgia to cope with economic hardships. According to the information obtained from the respondents, the monthly amount that was most frequently sent to their country of origin ranged from 4,000 to 8,000 rubles (€100-200). This amount was mentioned by 40% of the migrants from Armenia and by 30.7% of the migrants from Georgia. About a quarter of the polled migrants (24.4% from Armenia and 24.8% from Georgia respectively) used to send less than 4,000 RUB as monthly remittances. Some 14.4% of the interviewed migrants from Armenia and 17.8% of migrants from Georgia reported that they could afford sending home from 8,000 to 12,000 rubles (€200 to 300) every month. The majority of the interviewed migrants from Armenia (70%) and from Georgia (62.4%) used to send the money by bank transfers. The ArGeMi study revealed that the amount of remittances was basically determined by the level of migrant's earnings and not by his or her gender.

Table 9: Monthly amount (in rubles) of migrant remittances from Moscow to Armenia and Georgia (% of remitters)

Remittances	Migrants from Armenia			Migrants from Georgia		
	Total	Men	Wo-men	Total	Men	Wo-men
Less than 4 000 rubles	24.4	18.6	35.5	24.8	20.6	33.3
4 000 to 8 000 rubles	40.0	42.4	35.5	30.7	32.4	27.3
4 000 to 12 000 rubles	14.4	16.9	9.7	17.8	17.6	18.2
12 000 to 20 000 rubles	5.6	6.8	3.2	7.9	10.3	3.0
More than 20 000 rubles	4.4	3.4	6.5	5.9	5.9	6.1
No answer	11.1	11.9	9.7	12.9	13.2	12.1

The size of migrant households in Moscow is smaller than in their countries of origin. According to the ArGeMi data, the overwhelming majority of households consist of one to four persons, but most typically of three-four persons. This is the case with 53.5% of the migrants from Armenia and with 40.5% of the migrants from Georgia. The share of single households is somewhat higher among the migrants from Georgians. On average, every third migrant from Georgia and every fifth migrant from Armenia have such households.

When answering the question about their current status of employment, six of ten migrants from Armenia and seven of ten from Georgia mentioned that they were employed. About three of ten migrants from Armenia and two of ten migrants from Georgia replied that they were students at universities or colleges. There were more unemployed among the female migrants from Armenia than among the male migrants from the country. Some 12.4% % of the female migrants from Armenia were housewives. The situation of migrants from Georgia was reverse since only 6.5% of female migrants from the country were housewives. This difference can be explained by the fact that the migrants from Georgia more frequently stay temporarily in Moscow while the migrants from Armenia are predominantly permanent residents of Moscow. There were children in one third of the polled migrant households. Some 27.5% of the interviewees from Armenia and 22% of those from Georgia had one child; 9% and 11% respectively have two children. There were pensioners in every tenth migrant household and adult unemployed persons in more than one third of the households.

The ArGeMi survey further revealed that 54.5% of the migrants from Armenia and 44.6% of those from Georgia did not have a job-related qualification or experience in their country of origin. First of all, this applied to young migrants. According to the self-assessment of the respondents, they all had a good command of their native languages and nearly all of them reported a good command of Russian. The latter point is rather important for their adaptation to the Moscow community.

The experts who were interviewed in the framework of the ArGeMi project were requested to make an assessment of the gender composition of the migrants who arrived to Moscow in 2008. According to the interviewees, the share of females among the migrants from Armenia in this year comprised from 30 to 55% and among the migrants from Georgia from 25 to 60%. One expert expressed the opinion that the female share depends on the type of migration flows. He estimated that on average 55-60% of all immigrants from Armenia and Georgia to Moscow were women, while their share among labor migrants was 30%. In previous studies it was registered that the share of able-bodied men among migrants arriving in Moscow comprised 70-85% (Drobiževa, Arutyunyan and Kuznetsov 2007). The estimates of the interviewed experts indicated a decline of the gender disparity and probably the feminization of the migration flows from Armenia and Georgia. One may assume that due to the nature of their employment women are not always and fully represented in the official statistics as well as in the results of the field studies. Informal employments in the sector of services, including households and the entertainment industry often remain invisible. The interviewed experts believed that in the longer term the female share in migration flows will further increase. The share of females among the migrants from Armenia is slightly higher than among those from Georgia.

The settlement patterns of migrants from Armenia and Georgia in Moscow have some spatial specifics. The ArGeMi survey again confirmed the results of previous studies conducted by the *Institute of Social and Political Studies* and by the *Center of Demography and Human Ecology of the Institute for Population and Economic Forecasting*, both of the *Russian Academy of Sciences*. In fact, migrants from Armenia and Georgia most often select for their residence the outskirts of the Northern Administrative District (Kurkino, Strogino, Northern Tušino, Mitino), the townships of the North-Eastern Administrative Districts (Otryadnoye, Northern Yaroslavskiy, Rostokino, Mar'ina Rošča, Losinnostrovskiy, Sviblovo, Babuškinskiy) and the townships of the South-Western Administrative District (Akademicheskiy, Zyuzino, Yasenevo, Gagarinskiy). Migrants from Armenia usually reside in the Southern Administrative District townships of Nagornyj, Eastern Bibirevo, Čertanovo, in the Western Administrative District Očakovo, Novoperedelkino, Ramenkiy, Fili-Davydkovo and Fili Park as well as in the Zelenogradskiy Administrative district (Rybakovskiy 2006: 146-147). This pattern is determined by the availability of relatively inexpensive housing of medium or low quality in these townships. The filters impeding the settlement of migrants in the more prestigious districts of Moscow are the social and professional structure of the population, the reputation of an address and also the social prejudices of Muscovites. According to the data provided by other research teams, the successful natives of Armenia can be found in districts like Zamoskvorečie, Yakimanka, Tverskoy and Prospekt Mira. There are many immigrants from Armenia in the prestigious areas of Sokol and Airport. These addresses indicate the material well-being of their residents.

The ArGeMi survey paid special attention to potentials for the return of migrants from Armenia and Georgia to their home countries. According to the majority of experts, every year an insignificant number of migrants who arrived in Moscow three and more years ago return to their countries of origin. The key reasons that impede the longer migration were described as follows:

- Economic reasons (unemployment in Moscow, completion or loss of an occupation, failures in finding or retaining an occupation);
- Administrative (inability to legalize the residence in Moscow);
- Familial circumstances (the need to stay with the family and to support it in the home country);
- The aims of the migration to Moscow were achieved;
- The expectations for successful settlement in the Russian Federation had failed;
- Problems with laws, violations of the migration law;
- High living expenses in Moscow;
- Xenophobia and corruption in Moscow.

Only two experts expressed the opinion that the real number of returnees was rather big, due to transnational migration. According to these experts migrants often live "in two countries simultaneously". This is an opinion that is supported by many migration researchers as an increasing trend in international migration. However, this trend does not necessarily motivate cross-boundary migrants to return to their countries of origin.

The estimated figures of returnees to Georgia are slightly higher than the estimated numbers of returnees to Armenia. Political tensions between Georgia and the Russian Federation, the imposition of a visa regime for the entry of citizens of Georgia to the Russian Federation serve as arguments for this assumption.

According to the interviewed experts, some migrants from Armenia and Georgia arrive in Moscow with the intention to migrate to third countries. These transit-migrants prefer the most developed countries as their destination, in particular the USA, Canada, Israel, Australia, members of the European Union (Germany, Greece, France, Bulgaria, the Czech Republic), but also the Near East (Lebanon, Iran). The experts believed that the destination choice is determined by the existence of an Armenian or Georgian community in those countries. One expert specified the figure of migrants who leave for other European countries, the USA or Australia with approximately 15-17% of all migrants from Armenia residing in Moscow. The experts noted that transit-migration from Georgia occurs less frequent, with the USA, Germany, Canada, Turkey, Israel, Greece, France, Lebanon, and Iran as favorite destinations.

Most of the interviewed experts did not expect any substantial change in the migration flows from Armenia and Georgia, on the one hand, and from Moscow, on the other. The most frequent answer in 2009 was "the situation will remain be much the same". With regard to migration from Armenia the experts noted that the country's migration potential is already exhausted. The result is the stabilization or 'freezing' of the current situation. The assumption was expressed that the level of education of future migrants to Moscow would rise and more young people would come to the Russian Federation in order to receive education there. Most of them will probably stay in Russia. But there was also the opinion that the influx of irregular migrants to Moscow may increase due to the continuing economic crisis in Armenia. The expectations concerning the migration exchange between the Russian Federation and Georgia were somewhat different since there exists a visa regime for citizen of Georgia and the international relations between the two countries are not settled yet. However, there are also similarities between the prospects of out-migration from Armenia and Georgia since the out-migration flows from Georgia have stabilized and the remaining migration potential of Georgia is limited, too. Due to these factors some experts foresaw a decline of the migration from Georgia to Moscow. The interviewed experts pointed out that the migration movement towards Moscow might increase if:

- The economic situation in Moscow would improve fast and the number of available jobs and the level of wages would rise accordingly ;
- Friendly or at least neutral relations between the sending countries and the Russian Federation would develop together with an improvement of the migration policies and the international cooperation;
- The economic lag between Armenia and Georgia, on the one side, and the Russian Federation would deepen and political conflicts in the South Caucasus would re-emerge.

The results of the ArGeMi survey on migrants from Armenia and Georgia in Moscow basically correspond to such expert assessments and expectations. 72.5% of the migrants from Armenia and 66% of those from Georgia expressed doubt that the majority of their compatriots would return to their countries of origin. When asked about their own intentions, only 27.5% of the respondents from Armenia clearly expressed their readiness to return to the home country. However, this applied to 40.5% of the migrants from Georgia. But the majority of the 'return' inclined respondents understood the question rather literally as referring to a trip back home. Most probably, the "returnees" will re-migrate to their work or studies in Moscow once their holidays or vacations in Armenia/Georgia would be over.

When asked about their preferences of migration destinations, the respondents selected Moscow mainly for the following four reasons:

- Wide range of opportunities for work and business;
- High level of wages and salaries;
- Support of friends of relatives;
- Good own command of the Russian language.

However, if the choice of destination would depend only on the respondents' decision, then most migrants from Armenia would opt for the USA (18.5%) or other countries (16%). Only 11% of them would choose Russia again. Most migrants from Georgia named Russia (15.5%), the USA (15.5%) and other countries (12.5%). But the percentages only refer to the respondents who answered the question. Half of the ArGeMi respondents (47.5% of the respondents from Armenia and 51% of those from Georgia) did not provide a definite answer to the question.

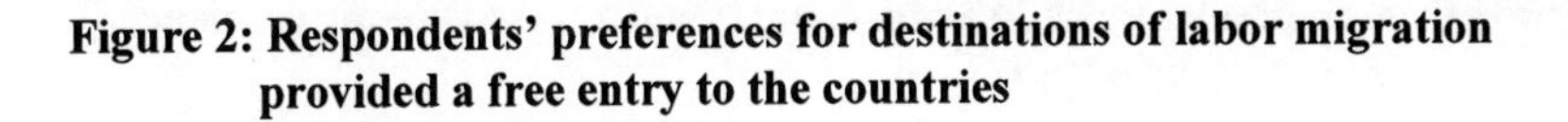

Figure 2: Respondents' preferences for destinations of labor migration provided a free entry to the countries

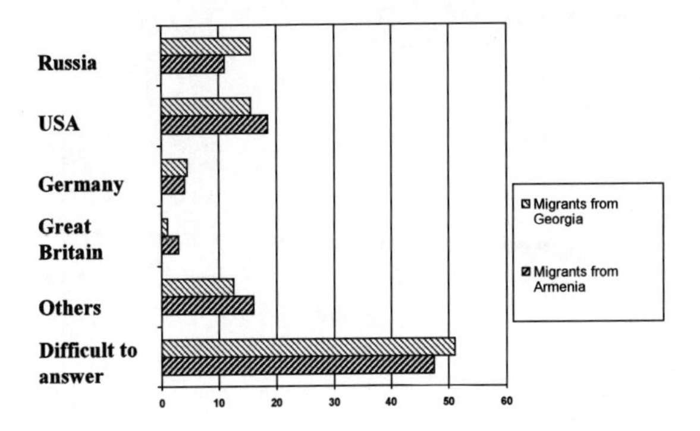

The preferences of desirable destination for residence – provided a free entry is granted – were nearly the same. Some 15% of Armenian migrants would prefer the USA, 29% would opt for other countries, and 9% of respondents would choose Russia. The respondents from Georgia would opt for Russia (12%), for the USA (9.5%), and 4% for Germany. Again, a high percentage of the respondents - 40.9% of the migrants from Armenia and 51% of the migrants from Georgia - found it difficult to answer the question. One of the major reasons may be the low share of those émigrés from Armenia and Georgia who reported a good command of the English language. According to their self-assessment, six of ten respondents from Armenia and seven of ten migrants from Georgia "have no command of English or have a poor command of that language".

Figure 3: A respondent's choice of a country most favorable for residence, free entry provided

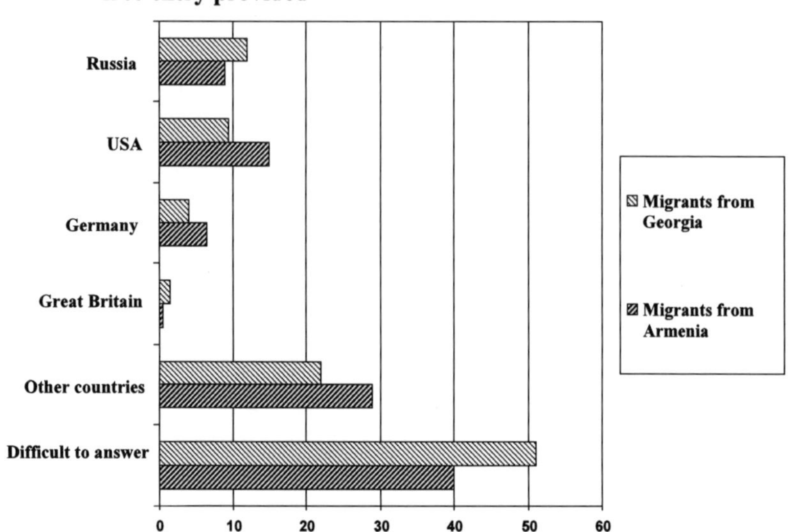

The ArGeMi respondents clearly differentiated between the alleged benefits from out-migration for their countries of origin and for the migrating individuals. Some 27.5% of the respondents from Armenia and 29% of the respondents from Georgia fully and respectively 39.5% and 43.5% partly agreed with the assumption underlying the following question: "Do you believe that out-migration is a blessing for Armenian and Georgian people – they can travel and work abroad freely?" Only a minority partly (8.5% and 11%) and respectively fully (6% and 3.5%) disagreed with this assumption. On the contrary, smaller shares of the respondents fully (7% of the respondents from Armenia and 9% of those from Georgia) or partly (28.5% and 35% respectively) agreed with the assumption "that out-migration is a blessing" for their countries of origin, because "it relieves the labor market and brings money into the country". Some 13.5% of the respondents from Armenia and 20.5% of the respondents from Georgia fully disagreed, while 18.5% and 13.5% respectively disagreed partly. Remarkably, nearly one third of the respondents from Armenia (32.5%) and more than a fifth of those from Georgia (22%) was not sure about their assessment of the situation.

In a different context, the balance of the consent (among the migrants from Armenia 12% fully and 27% partly, from Georgia 10% and 28% respectively) and dissent (migrants from Armenia 20.5% fully and 14.5% partly, from Georgia 15% fully and 21.5% partly) with the guiding assumption of the question "Do you believe that out-migration is a curse for the Armenian/Georgian people – migrants suffer exploitation and discrimination abroad, families suffer as well?" is approximately the same. The majority of the respondents also agrees with the statement that Armenia and Georgia lose their "best citizens" due to migration (consent among the migrants from Armenia 35% fully, 27% partly, from Georgia 26.5% fully and 36% partly).

5 The Integration of Migrants from Armenia and Georgia in the Moscow Megalopolis

The movement of migrants from Armenia and Georgia to the huge and unfamiliar city of Moscow is unavoidably full of tensions. They concern the settlement at a new place, the overcoming of the cultural shock, which is experienced by migrants, the formation of a new identity which is acceptable for migrants as well as for the local population. The ArGeMi field study revealed that the migrants from Armenia and Georgia are striving to secure a high quality of habitat for themselves and their families. However, the migrants are confronted with serious practical tasks like search for or acquisition of accommodation, employment, legalization of residence, shortage of information and discriminatory behavior of authorities.

Moscow is one of the most expensive cities in the world. Prices for the acquisition or renting of accommodation are very high. Even when prices for real estates went down in 2009, they still remained high. One has to work several months to accumulate a sum sufficient for the acquisition of just one square meter of a good housing. Rentals are also rather high. The rental for a two-room apartment depends on the township and ranges from 28 thousand rubles to 70 thousand rubles (from €700 to 1700) per month. Under these circumstances the ArGeMi respondents (44% of the migrants from Armenia and 48.5% of migrants from Georgia) most frequently live in rented apartments or rooms. A quarter lives in hostels and boarding-houses or with relatives or friends. Migrants from Armenia stay more often "with friends and relatives" than the migrants from Georgia. The migrants who have already obtained the citizenship of the Russian Federation and have acquired accommodation in Moscow make out the majority of the respondents from Armenia. The majority of the respondents from Georgia who have bought an apartment in Moscow also consist of persons who have obtained the Russian citizenship. Nine of ten respondents are generally satisfied with their accommodation. Half of the interviewed migrants from Armenia and 40% of the migrants from Georgia were fully satisfied with their housing conditions in Moscow. Only 7% of the respondents are dissatisfied. Women, nationals only of Armenia and Georgia and persons older than 35 years used to be more often dissatisfied with their housing in Moscow than the naturalized or male cohorts.

The Moscow labor market has developed patterns of migrants' employment which are typical for recipient countries. As a rule, the migrants from Armenia and Georgia are employed in spheres that are unattractive for Muscovites. The ArGeMi respondents were predominantly employed in the retail trade. The next frequent employments for migrants from Armenia are construction, the public health system, and transportation. For the migrants from Georgia the employment in the construction, transportation, and the education system follow the employment in the retail trade. A significant share of the respondents (19% of the employed migrants from Armenia and 15.3% of the employed migrants from Georgia) indicated the "other sphere" as an answer to the question concerning their employment. 7.7% of all respondents from Armenia and 16.3% of those from Georgia who have established own businesses proved to be the most active cohort of migrants working in Moscow.

Most experts mentioned the same occupational sectors for migrants from Armenia and Georgia as they were indicated by the interviewed migrants themselves. The experts also noted that migrants from Armenia were employed in science, culture, housing and utilities, too, and run own small and medium businesses. According to the experts, the migrants from Georgia were also involved in private transportation services, restaurant and gambling business, cultural, entertainment and leisure business. It was repeatedly mentioned by the experts that in comparison with migrants from Central Asia, the migrants from Armenia and Georgia are better educated and the range of their employments is wider. One expert called the migrants from Armenia and Georgia "the elite of labor migration". When characterizing the migrants from Armenia in particular, the experts specially pointed out to their high level of education, adaptability to new or changing conditions together with their zeal to integrate into the Moscow community and their abilities to develop serious business. Some experts expressed the opinion that employers prefer migrants from Armenia because they are believed to be "workaholics" and good specialists. Construction firms are particularly interested in migrants from Armenia because they "have a knack for making things work" and are law-abiding. But one expert noted that some employers give preference to migrants from Central Asia because it is cheaper to hire them than migrants from Armenia or Georgia.

The ArGeMi survey revealed that migrants from Armenia and Georgia chose those sectors of the Moscow economy that more or less correspond to their previous work experience. But the migrants are also ready to make some concessions in this respect in order to adapt to the requirements of the labor market in Moscow. The share of the migrants from both countries who are employed in the industry, in the spheres of education and in agriculture is lower than the work experience of the migrants in their country of origin might suggest. The finding might be interpreted as a surprise that 14.4 per cent of the migrants from Georgia had work experience in the construction industry in their home country and 13% of the migrants from this country are employed in the construction industry in Moscow. The same striking results concerned the employment in the trade sector since 33.4% of the migrants from Georgia had work experience in this sector in Georgia and 38.9% of the migrants from Georgia were employed in the trade in Moscow. The same concerns the experience of

26.5% of the migrants from Armenia who had work experience in trade already in their home country. In Moscow 31.9% of the interviewed migrants from Armenia were employed in the trade. The share of men employed in traditionally 'male' economic sectors of physically hard work such as construction or transportation is higher than the share of women. On their turn, they prevail in the trade, education, and public health.

Table 10: Distribution of the migrants from Armenia and Georgia in the economic sectors in their countries of origin and in Moscow (% of the employed migrants)

Branch	Migrants from Armenia				Migrants from Georgia			
	Armenia	Moscow			Georgia	Moscow		
		Total	Men	Women		Total	Men	Women
Industry	11.0	6.9	10.8	2.2	13.5	6.9	8.9	3.7
Construction	8.8	14.7	20.3	4.6	14.4	13.0	19.0	3.7
Transportation	11.0	9.5	13.5	2.2	12.6	11.5	15.2	5.6
Agriculture	7.7	0.9	1.4	0.0	4.5	0.0	0.0	0.0
Trade	26.4	31.9	25.7	41.3	33.4	38.9	36.7	42.6
Education	14.3	6.9	2.7	15.2	13.5	9.2	5.1	14.8
Public health	12.0	10.3	5.4	19.6	3.6	5.3	0.8	7.4
Other	8.8	19.0	20.3	15.2	4.5	15.3	11.4	22.2

The ArGeMi field study confirms the repeatedly registered fact that migrants from Armenia and Georgia mostly resort to unofficial channels and informal ties in their search for employment. Their occupational choice or choice of business activities are also determined by the support provided by friends and relatives. Every second respondent who tried to get employment or to start his/her own business resorted to the help of friends and relatives. Less than one percent of the respondents from Armenia and 1.6% of those from Georgia got employment or started their businesses by using the support of employment agencies in their country of origin. Some 7.8% and 11.4% respectively did so in Moscow. A significant 36.5% of the respondents from Armenia and 38.2% of those from Georgia found employment or started their own business by exclusively relying on their own initiative. Obviously, there are deficiencies in the functioning of the employment system for migrants in Moscow and even more so in the countries of origin. The institutions of kin relationships and friendship remain crucially important for the majority of migrants looking for jobs in the capital city of the Russian Federation. The ArGeMi survey furthermore delivered the information that the ethnic origin of the employer was not crucially decisive in the respondents' search for employment. More than half of the respondents were employed in enterprises owned or managed by nationals of the Russian Federation. Nevertheless, it turned out that every tenth respondent from Armenia and Georgia was employed by a compatriot.

The ArGeMi survey also indicated that 'brain waste' was a real issue for the interviewed migrants. More than a quarter of the respondents from both countries complained that their current employment or business in Moscow did "not at all" correspond with their level of education. A partial discrepancy between the respondents' education and their employment was observed in about 40% of all cases. The 'brain waste' issue seemed to be particularly urgent among the women from Georgia. Some 39% of them answered the question concerning the correspondence between their education and type of occupation with "not at all". In the sample of the migrants from Armenia the lack of correspondence was reported by 26.2% of the men and 26% of the women. One should notice that approximately the same proportions of correspondence between the type and level of education and the educational requirements of employment are typical for the 'old' residents of Moscow, too (Dynamics of the social sphere of Russia 2009: 50). The discrepancy partly explains by the dynamic changes on the capital city's labor market.

The changes of work profiles and other difficulties notwithstanding, the overwhelming majority (9 of 10) of the employed ArGeMi respondents were basically satisfied with their current employment. One third of the migrants from Armenia and one quarter of those from Georgia expressed complete satisfaction in this respect. As a rule, female respondents and nationals of Armenia and Georgia were less satisfied with their employment situation. One explanation for the dissatisfaction of the nationals of Armenia and Georgia might be related to the higher probability of an irregular status among them.

Table 11: Level of satisfaction with the employment or business in Moscow (% of the employed)

	Migrants from Armenia			Migrants from Georgia		
	All migrants	Males	Females	All migrants	Males	Females
Completely satisfied	33.0	27.7	40.0	25.2	22.9	28.3
Partially satisfied	58.3	66.2	48.0	62.6	68.6	54.7
Dissatisfied	8.7	6.2	12.9	12.2	8.6	17.0

The economic crisis reduced the demand for migrants' labor in Moscow. According to data of the *Department of Labor and Employment of the Moscow population*, at the time when the ArGeMi survey was conducted in 2009, 74.5 thousand persons were unemployed and actively looking for employment. Of these 56.9 thousand persons were officially recognized as the unemployed (Rynok truda 2009). The Government of Moscow responded to the deterioration of the social and economic situation by halving the quota for work permits to foreign workers in 2009 (On approval of the quota….2008). In 2010, the quotas were again reduced by a factor of 1.5. The intention of such reduction was the authorities' desire to protect the right to labor of the local employees and to prevent the presence of unemployed migrants in Moscow. Given the distribution of spheres of employment between migrants and Muscovites, the authorities could only minimally reduce the influx of migrants. The measures taken by authorities rather enlarged the shadow employment. Despite the reduction of construction activities, wholesale and retail trade, they still occupied the top positions in the local search for labor force in Moscow in 2009. Both are branches of predominant employment of migrants from Armenia and Georgia. A reduced level of formalization of migrant employment certainly has negative impacts on their social protection. According to data provided by E. Tyuryukanova, the migrants employed in Moscow's informal labor market are typically underpaid for their work. Overtime work is not always paid, wages are often detained. Migrants are forced to perform extra functions or to provide sexual services. Informally employed migrants are often unjustly sacked (Tyuryukanova 2006).

The employment of migrants from Armenia and Georgia provides them and their families with relatively high earnings that are comparable with the average per capita incomes of Muscovites. According to data provided by the *Moscow Statistical Office,* in the period from January through March 2009 the average per capita money income of the employed in Moscow comprised 34,814 rubles (Osnovnye pokazateli denežnykh dokhodov...2009), while the reported average per capita income of employed migrants from Armenia comprised 35,652 rubles and the average per capita income of employed migrants from Georgia comprised 31,263 rubles. Every fifth respondent had a monthly income in the 24,000-40,000 RUB range. Every tenth migrant from Armenia and some more than every tenth of those from Georgia mentioned average monthly incomes of more than 40,000 RUB. This personal income of the respondents has to be related to the number of their household members and the income situation of all household members. It has to be recalled that 53.5% of the respondents from Armenia and 40.5% of those from Georgia live in households with three or four members. Then each member of these households has an estimated monthly income of 8,500-11,333 RUB, which is a three times lower then the per capita income estimated by the Moscow Statistical Office for the population in Moscow.

The respondents did not indicate the health care for migrants in Moscow as a particularly significant issue. Only 7.5% of the respondents from Armenia and 11.5% of those from Georgia reported difficulties with health care in Moscow. However, only 59% of the respondents from Armenia had a permanent health insurance, while 23% were temporarily insured. The respective data for the respondents from Georgia were lower (47% and 27%). Every fifth migrant from Armenia and every fourth migrants from Georgia did not have a health insurance at all. That may deprive them of health care in case of need. The self-treatment could cause inability to work or bring about chronic disease. Young bachelors, who are not citizens of Russia, predominate among those respondents from Armenia who are not insured. This largely applies to respondents of all ages from Georgia, who have no Russian citizenship. The segment of the uninsured respondents was mostly composed of migrants who have recently arrived to Moscow or temporary migrants, who do not intend to stay in Moscow for a longer period.

The survey had a special focus on the communicative and cultural practices of the migrants from Armenia and Georgia. The findings demonstrated that while being in Moscow the overwhelming majority of respondents (85.5% of the migrants from Armenia and 84.5% from Georgia) kept contacts with relatives and friends living in their country of origin.

Table 12: Contacts of migrants from Armenia and Georgia with relatives and friends residing in their country of origin (% of respondents)

Frequency	Frequency of phone calls		Frequency of Internet communication	
	Migrants from Armenia	Migrants from Georgia	Migrants from Armenia	Migrants from Georgia
Every day	9.0	8.5	20.5	20.0
Every week	38.5	42.5	20.5	19.0
Every month	20.0	19.5	9.0	12.5
Once in several months	18.0	14.7	5.5	11.5
Not at all	14.5	15.5	44.5	37.0

More than a half of the respondents host relatives and friends as guests no less than once a year. Every second migrant visits his/her homeland at least annually. This allows concluding that the majority of migrants from Armenia and Georgia retain close ties with their country of origin.

Table 13: Frequency of visits to a respondent made by friends and relatives from a country of migrant's origin (% of respondents)

Frequency	Migrants from Armenia	Migrants from Georgia
Every month	3.0	9.0
Twice a year	11.0	15.0
Once a year	45.0	29.5
Not at all	37.0	45.0
Other	4.0	1.5

However, the analysis of migrants' cultural practices also revealed their absorption in the Russian information context. As a rule, during their stay in Moscow migrants read Russian printed media and watch mainly Russian TV. Only a third or perhaps a quarter of migrants focuses on media from their countries of origin once a week or once a month. One may assume that relatives and friends with whom the migrants keep contact are another important information source about events in Armenia.

The majority of migrants who are naturalized and have families in Moscow do not visit ethnic clubs or organizations in the capital city. However, every fourth ArGeMi respondent from Armenia and 16.5% of the respondents from Georgia used to attend compatriots' club or organization in Moscow once a month. Merely one percent of the respondents from Armenia and 2% of those from Georgia attend such institutions more frequently. Young bachelors are the most active participants of club activities.

The analysis of the social contacts bears witness to the respondents' successful integration into the Moscow community. The overwhelming majority (57.4% of the migrants from Armenia and 65% from Georgia) mentioned that their colleagues and business partners treated them "friendly". 27.8% of the respondents from Armenia and 17.1% of those from Georgia qualified the relationship as "very friendly". The relationships with colleagues at the work place or in the business environment were qualified as "indifferent" by 13% of the migrants from Armenia and by 16.3% of those from Georgia. Only a negligible 1.7% of the migrants from Armenia and 1.6% of those from Georgia reported about unfriendly attitudes by colleagues and business partners. Not a single respondent assessed the attitude of colleagues and partners as 'hostile'. When asked about the national background of their colleagues and business partners, the respondents from Armenia and Georgia mentioned in the first place nationals of the Russian Federation (63.5%of the respondents from Armenia and 63.4% of those from Georgia). Next to citizens of the Russian Federation as colleagues and business partners of the respondents came their compatriots (21.7% for the migrants from Armenia and 18.7% for the migrants from Georgia), but also migrants from third countries (14.8% and 17.9% respectively).

A similar ranking could be identified in the national affiliation of personal friends. Russians were most frequently mentioned followed by compatriots and expats:

Table 14: Friends of respondents (multiple choice, % of respondents)

Friends of respondents	Migrants from Armenia			Migrants from Georgia		
	All	Males	Females	All	Males	Females
Russians	58.0	45.9	73.0	44.5	46.3	41.6
Fellow nationals from the country of origin who arrived to Moscow after 1990	39.0	44.1	32.6	37.0	37.4	36.4
Representatives of the country of origin residing in Russia prior to 1990	17.5	16.2	19.1	15.0	16.3	13.0
Representatives of other nationalities	12.5	15.3	9.0	19.5	19.5	19.5
No friends in Moscow	3.0	5.4	0.0	3.5	2.4	5.2

Similar to the preferences in the respondents' social life, their behavior in crisis situations is not primarily oriented towards compatriots or ethnic networks. In cases of crisis or emergencies respondents usually resort to:

- Their local acquaintances (52% of the respondents from Armenia and 55.5% of those from Georgia);
- Compatriots from Armenia or Georgia (38% of the respondents from Armenia and 46.5% of those from Georgia);
- Local authorities (15% of the respondents from Armenia and 14.5% of those from Georgia);
- To their national embassy (3.5% of the respondents from Armenia and 3.5% 10% of those from Georgia);
- A lawyer (7.5% of the respondents from Armenia and 8.5% of those from Georgia);
- To human and civic rights NGOs (5% of the respondents from Armenia and 2% of those from Georgia)
- Other organizations (2% of the respondents from Armenia and 5.5% of those from Georgia).

Some six percent of the respondents from Georgia replied that "one cannot expect help in Moscow" at all. A surprisingly insignificant, even negligible share of respondents would recourse to their national Church representations (one percent of the respondents from Armenia and two percent of those from Georgia). This rather restricted choice can be explained by the development of integrative cultural ties and the formation of wider social networks. They include colleagues and neighbors who are selected not on an ethnic basis but on the basis of mutual sympathy and personal commonalities. Most often 'old' Muscovites work side by side with the migrants. The locals possess the needed social capital due to their establishment in the capital community. This is why the resort to such persons meets the migrant's expectation that they would get a support.

Table 15: Persons or institutions to which migrants apply for a help or advice in case of emergency in Moscow (multiple choice)

Approached persons or institutions	Migrants from Armenia		Migrants from Georgia	
	%	Standing	%	Standing
Fellow nationals from Armenia or Georgia	38.0	2	46.5	2
Local acquaintances	52.0	1	55.5	1
Embassy	3.5	8-10	10.0	4
Local authorities	15.0	3	14.5	3
A national migrant organization	3.5	8-10	2.5	9-10
Representatives of national Church	1.0	12	2.0	11-12
Human rights organizations	5.0	7	2.0	11-12
Other organizations	2.0	11	5.5	7-8
Lawyer	7.5	5	8.5	5
Other	6.5	6	5.5	7-8
No one in Moscow will help	3.5	8-10	6.0	6
No answer	12.0	4	2.5	9-10

Due to the specifics of mentality, shared values, traditions and patterns of conduct, the majority of the respondents (71% of the migrants from Armenia and 65.8% of the migrants from Georgia) preferred to marry a member of their ethnic group. Only a third of the respondents from Armenia and a quarter of those from Georgia were married to ethnic Russians. 7.9% of the respondents from Georgia had spouses, who belong to other ethnic groups.

When assessing the attitude of the locals to them, the respondents made a clear distinction between the representatives of the authorities and the general population. Nearly two thirds of the respondents from Armenia and half of those from Georgia described the attitude of the 'average people' in Moscow as 'friendly', or "very friendly", while only 2.5% and 7.5% respectively mentioned an 'unfriendly' attitude. Despite this positive evaluation, 11% of the respondents from Armenia and 26% of those from Georgia felt insecure during their stay in Moscow. When asked about their experience of discomfort, about one third of all respondents reported verbal offences. 8.5% of the respondents from Armenia and 5% of those from Georgia had been physically attacked. Migrants from Georgia explained these phenomena by the particular antipathy of Muscovites against Georgians (30.5%), foreigners (28.8%) and labor migrants (25%).

Similar results were obtained by the monitoring on Russia's social sphere and the examination of the way of life of Russians in the post-Soviet space in 2008. A half of the Muscovites were opposed to the increasing number of workers from the CIS countries (Osadchaya 2009: 36). They assumed that the attraction of a considerable number of migrants would threaten the capital's security. Muscovites also expected that migrants should adjust to the recipient megalopolis culture. Negative attitudes were found, first of all, against the natives from the 'Caucasus' with no distinction between the various ethnic groups. The majority of the ArGeMi respondents assessed the attitude of the local authorities in Moscow or of the Federal authorities as indifferent. This was the assessment of 55.5% of the migrants from Armenia and 57.5% of those from Georgia. However, 12% of the respondents from Armenia and 17.5% of those from Georgia described this attitude as "unfriendly" or "hostile".

The interviewed experts noted that the authorities usually do not consider the migrants from Armenia and Georgia as specific groups. Nevertheless, in the experts' opinion the migrants from Armenia are typically perceived and treated more positively while migrants from Georgia are perceived as a group of enhanced risk and as undesirable in Moscow. Thus, if the authorities' attitude towards the migrants from Armenia is indifferent or reserved their attitude towards migrants from Georgia is mostly cautious or implies more negative evaluation or treatment. In addition, the overwhelming majority of experts noted that migrants from Armenia and Georgia enjoy formal rights prescribed by laws but do not enjoy real rights in practice. However, this ambiguity applies not only to migrants from the South Caucasus, but to migrants from other countries as well. The experts noted in addition that the same holds true for the locals in Moscow as well. They also have problems with the police. The experts acknowledged the necessity for certain measures to integrate migrants. The information system and the legal support should be improved, programs of legalization should be developed, labor and social guarantees should be secured, and education of children should be provided. The *Federal Migration Service* should cooperate with migrant communities and organizations in the prevention of irregular stays.

Despite all the above mentioned difficulties, 40.5% of the respondents from Armenia considered their current stay in Moscow as 'entirely satisfactory'. The same applied to 42% of the migrants from Georgia. In addition, 54% of the migrants from Armenia defined their experiences from the stay as 'partly' satisfactory. The same applied to 56.8% of the migrants from Georgia. 'Entirely satisfied' were mostly those respondents who had received the Russian citizenship and had families. About 5% of all respondents were dissatisfied with their stay in Moscow. This group consisted mostly of nationals of Armenia and Georgia older than 35 years. One may assume that they had higher expectations which they could not satisfy.

6 Regulation of the Migration Processes in the Moscow Megalopolis

The Russian and Moscow migration policy had gone through a series of changes during the last two decades. One of the interviewed experts identified five stages of migration policy transformation. The first stage covers the migration policy from 1991 to August of 1994. This stage consisted of reactions to the challenges of forced massive migration. Under the conditions of the ruined system of management and a legal vacuum Russia was for the first time confronted with the problem of massive forced immigration. The number of migrants who arrived to Russia in the early 1990s from the former Soviet republics was not much different from the figure of arrivals since the mid-1980s. But it was precisely the problem of refugees and forced migrants which defined the new situation and the respective laws. The second stage covered the period from August 1994 to May 2000. This stage was characterized by the efforts to specify the goals and tasks of the migration policy. The existing laws were rectified to be adapted to the needs of the forced migrants and refugees. Two basic laws (*On forced migrants* in the wording of the Federal Law # 202 of December 20, 1995, and *On refugees* in the wording of the Federal Law # 95 of June 28, 1997) underwent important amendments. Both were significant breakthroughs in comparison to the laws passed in 1993. Amendments specifying the passport of the citizen of the Russian Federation and the documents needed for the transit through the territory of the Federation were introduced in the Law *On the procedure of entry to the Russian Federation and departure from the Russian Federation*. In 1995 a decline of immigration began. One of the reasons was the inability of the state to accomplish the obligations toward migrants as they were enshrined in these laws.

The third stage encompassed the period from May 2000 to February 2002. It was the period of the reorganization of the Federal Migration service and the transition to a heavy-handed approach to migration issues. According to the experts, the functioning of the migration service was paralyzed during these two years due to reorganizations that brought about a principal change of priorities in its activities. The fourth stage encompassed the period from February 2002 to January 2007. It was the time of rigorous deliberalization with the aim of "closing the country". In February 2002 the *Federal Migration Service* became part of a law-enforcement agency, with issues of registration and control as the FMS's priorities. One of its main activities was the struggle against irregular migration.

The fifth stage covers the period from January 2007 to April 2009. It was the period of partial re-liberalization. A new Federal Law *On the Registration of Foreign Citizens and Stateless Persons in the Russian Federation* was issued. The registration was previously permissive but was further changed to the notification type registration. Radical amendments were incorporated in the Federal Law *On the Legal Status of Foreign Citizens in the Russian Federation.* These amendments considerably facilitated the employment of migrant workers from CIS countries. However, the expert noted that attempts to fully implement liberal approaches had failed. Consequently, the migration policy at the time of the ArGeMi field studies (the first half of 2009) was characterized by a plentitude of internal contradictions which were aggravated by the global financial and economic crisis.

Another expert complemented the above periodization of the RF migration policy by distinguishing an additional stage from the beginning of 2009 on. The expert defined this stage as marked by quota curtailments and the growth of irregular migration. In general terms, the experts assessed the migration policy in Moscow and in Russia as short-sighted and lacking a clearly articulated strategy. As far as it exists and functions, the migration policy of the Russian Federation was defined as inefficient in terms of the protection of the national interests. Its main goals were and remain the control over the activities of enterprises and the imposition of fines on irregular migrants. The Russian laws and the migration policy of the Russian Federation were also described as not fully adequate and not fully adapted to the international experience and requirements. It is not by chance that the Russian Federation still had problems with signing the *UN* and *ILO Conventions* because the Federation was not ready to fulfill the legal and even less the practical requirements. The experts used to note that these requirements could serve as the basis for the development of effective laws and interstate relations. Nevertheless, several experts added that the attention to migration issues, particularly to irregular migration issues was rising in Russia.

The opening of a *Center for Provision of Services to Legal Entities and Individuals* on issues of labor migration in Moscow was often quoted as an example of this increased attention to the migration issues. The Center was established on the basis of the units of the *Directorate of the Federal Migration Service* in Moscow. The establishment of similar centers all over Russia was under consideration. The new institution was to become a mechanism for preventing corruption at local levels and for full control over receipts of payments for state fees (Žebit 2009).

According to expert judgment, the Russian leaders have recognized the necessity of a more liberal migration policy in general and in the area of labor migration in particular. The elaboration of this policy is being carried on. Measures aimed at the integration of migrants had been undertaken. They include the teaching of Russian to migrants and the distribution of relevant information to them. The integration of newcomers should be supported by their compatriots already settled in the Russian Federation. Some experts criticized the nearly complete lack of attention paid to issues of the migrants' social integration into a new context. Most interviewed experts thought that there was practically no policy of migrants' integration, nor was there any understanding of who should be integrated and in which way integration should be pursued. There was no identifiable administrative body which was responsible for such efforts. One expert pointed out that the Moscow administration attempted to do something to this end, but its integration policy was based on obsolete approaches to migration.

Therefore, the majority of experts came to an ambiguous assessment of the migration policy of the Federal Government. They concluded that the government "promoted and hindered migration" simultaneously. Explaining their opinion, some experts noted that the Government used to encourage certain categories of migrants while it discouraged others. The experts also mentioned that the Government "encouraged migration prior to the crisis" and still "encourages migration of fellow nationals, but hinders temporary labor migration" and "encourages migration for permanent residence". Other experts argued that the Government did not hinder migration to the Russian Federation at all, provided the migration corresponded with the existing laws.

The judgment "hinders" prevailed in the experts' assessment of the migration policy of Moscow City Government. One expert qualified its policy toward migrants as restrictive, with increased taxes and decrease of the public expenses as main restrictive means. All these approaches were seen as opposed to the generally liberal policy of the Government of the Russian Federation. However, several other experts believed that the Moscow Government encouraged migration and did not hinder even the mass immigration of migrants from Tajikistan, provided the migration was legal.

All experts acknowledged the necessity of migration and its significance for the development of the capital city. Without migrant labor the megalopolis could not function. The city needs both skilled and unskilled migrants but it needs unskilled labor more, and migrants are ready to perform works that Muscovites refuse to perform (street cleaning, roadway maintenance work, etc.). The majority of the interviewees also believed that immigration always poses risks, first of all for the migrants themselves, because migration is a leap into the dark. The experts also noted the low level of tolerance in Moscow, the corruption among the law enforcement bodies and particularly the negative attitude towards natives from the Caucasus. A number of experts focused on the risks for the recipient country by mentioning security and a growth of crime first of all. Migration was also regarded as problematic since it brings about overpopulation in the city and increases interethnic tensions. According to these experts, semi-regular and irregular employment of migrants causes the greatest risk while regular migration is safe. The main ways to avoid risks is the improvement of the efficiency in the control and management of migration processes, and the implementation of a well designed integration policy.

According to the interviewed experts, the amended version of the law *On the legal status of foreign citizens* represented a groundbreaking piece of legislation since it brought about an impressive positive effect already during the first year of its implementation. In the national report *On the development of the human potential in the Russian Federation in 2008: Russia in the face of demographic challenges* (2009) for the first time relatively authentic data on the total figure of migrants were published. This contributed to the observation of the human and labor rights of the migrants. If prior to the amendment some 46% of all migrants were unregistered, the specific weight of the unregistered migrants decreased to 15% after the amendment. The overwhelming majority of migrants were registered and thus removed from the shadow economy. The processing of work permits was drastically improved. According to data of the ILO and OSCE monitoring in 2007, 75% of all labor migrants got work permits. In the previous times mere 15-25% of the migrants had been employed by employers who had respective permits. Yet it was noted in the same report that the new procedure of work permits issuance for migrants once again revealed the ambiguous nature of the Russian labor market and the availability of vast areas for shadow employment. About 40% of the migrants who obtained work permits were nevertheless employed by their employers in an informal way. The employers resort to this practice in order to escape the payment of taxes. Thus, an absolutely legal migrant who has fulfilled all formal legalization requirements nevertheless may turn out to be an irregular worker. Such migrants might be even unaware of their illegal condition.

The interviewed experts used to repeat that migration issues were regulated on the basis of bilateral agreements concluded between Russia and Armenia. The experts also pointed out at the most significant legislative acts governing the cooperation of Russia and Armenia on migration issues:

- The agreement on *Cooperation in the sphere of labor migration and social protection of labor migrants in the CIS countries.* The agreement was signed on 15 April 1994. Georgia entered the agreement, but in 2009 officially withdrew from the CIS.

- The recommendatory legislative act *Migration of labor resources in the CIS countries,* adopted by the CIS Inter-Parliamentary Assembly on 13 May 1995;
- The agreement *On cooperation in struggle against irregular migration,* signed on 6 March 1998;
- The *Declaration on a coordinated migration policy*, approved on 5 October 2007.

A convention on the legal status of labor migrants and members of their families was mentioned as a forthcoming step in the regulation of migration flows in the framework of CIS. The draft of the agreement was characterized as well based on provisions of international conventions. This new regulation will facilitate the inter-state interaction with Armenia on migration issues. The bilateral relations between the Russian Federation and Armenia were regarded by the experts as legally well regulated by the following documents (Dvustoronnye soglašeniya 2010):

- Treaty between the Russian Federation and the Republic of Armenia on the legal status of the RF citizens permanently residing in the territory of Republic of Armenia and the citizens of the Republic of Armenia permanently residing in the territory of the Russian Federation (29 August 1997);
- Protocol between the Government of the Russian Federation and the Government of the Republic of Armenia on the extension of the duration of the Agreement between the Government of the Russian Federation and the Government of the Republic of Armenia on the regulation of the voluntary resettlement process (29 August 1997).

According to the interviewed experts, bilateral agreements are more effective as instruments of governing specific issues of migrant protection than multilateral agreements. However, the legal regulations of migration are generally regarded as obsolete and inefficient. The existing legal regulations do not ensure the implementation of migrant rights since they only tentatively define the implementation procedures.

When asked whether the Federal and Moscow City authorities consider migrants from Armenia and Georgia in any special way or not experts held different views. The overwhelming majority believed that the authorities do not demonstrate any particular attitude towards migrants from these two countries. However, some experts believed that a component of political disagreement and a higher risk of crime were taken into consideration by authorities with regard to the migrants from Georgia.

As to the migration policy of Moscow City Government, it was reported to be based on a number of normative documents issued by the Moscow authorities:

- Decree of the Mayor of Moscow # 1090 *On the procedure of registration and the execution of foreign citizens working in the sphere of services of Moscow City* of 13 October 2000;
- Decree of the Mayor of Moscow # 447 *On the procedure of registration of foreign citizens working at the urban markets, fairs, in malls trade complexes, in strip malls and trade houses located in the territory of Moscow City* of May 2001;
- Resolution of the Moscow City government # 189 *On the approval of rules of registration and deregistration of the citizens of the Russian Federation at places of their staying and residences in Moscow City* of 6 April 2004.

Special mention was made by the experts of the *Program on cooperation in the economic and humanitarian spheres* which had been signed by the mayors of Moscow and Yerevan in 2005. It was underlined in the interviews that the document well defined the productive relations between the two capitals. The relevance of the *Program* exceeded the limits of the relations between both cities since it was signed in the framework of the *Treaty on friendship, cooperation and mutual assistance between the Russian Federation and the Republic of Armenia*.

The special migration program for the period from 2008 to 2010 was stressed as a comprehensive document fixing the major ideas of the Moscow City leaders about policy directions. The targets and tasks of the city's policy in the sphere of labor, social protection, social welfare and education of migrants, and their integration into the megalopolis community were precisely formulated in this *Program*. The idea of organizing the arrival of migrants on the principle of "single window" was included into the *Program* together with the suggestion to build up simple hostels. The migrants could register and reside in the hostels (Moscow City Program 2007). However, the *Program* had been implemented only as a pilot project and had exerted no significant impact on the overall migration situation in the city. Despite Moscow's obvious interest in migrants, the *Program* did not contain a detailed vision about the integration of migrants in the recipient community. The experts discovered even incompatibilities between some provisions of the *Program* and the Federal laws. Some examples concern the suggestion to register only those migrants who have concluded agreements of accommodation and to issue work permits only to those who have already found a job within the assigned quota. The idea about the hostels for migrants was also criticized because the hostels were suggested as administrative settlements for foreign workers. Migrants will most probably see them as an institution facilitating their alienation from the local population. The intention of the

Moscow City Government to reduce the number of foreign workers by the imposition of strict quotas as well as the dependency of a migrant on his/her employer were also regarded as ideas that demand reconsideration. The expert community did not support these aspirations of the Moscow City Government since they reveal the ambivalent nature of its migration policy. As Anton Paleev, a member of the *Moscow City Duma* admitted, migration provides for an influx of labor force in important sectors of economy like construction, transportation and trade. On the other hand, migration is regarded as disturbing the 'natural' balance in the labor market, causing a sanitary and epidemic threat and significantly increasing crime rates. Finally, he mentioned that migration tends to increase the load on the social sphere (polyclinics, kinder-gardens, schools, even on the drainage and water supply). In his interview Paleev pointed out that, according to data of the city public prosecutor's department, every third murder and every third rape were committed by migrants (Paleev 2009).

One of the key reasons for the lack of a clearly defined program of migrant integration into the Moscow City community was related by the experts to the fact that none of the existing models of integration (assimilation model, multicultural model) was accepted by the Moscow community. The Constitution of the Russian Federation (adopted in 1993) provides for the right to preserve native languages, including the free choice of a language of communication, instruction, education and creative activities (Article 26), the protection of the rights of smaller ethnic groups (Articles 68, 69) and provisions forbidding propaganda of social, racial, national or linguistic superiority (Article 29). The Federal Law *On National Cultural Autonomy* (adopted in 1996) is in conformity with the international law and international treaties. In it the rights of minorities and in particular the language rights were enshrined - beginning with the *Universal Declaration of Human Rights* (United Nations, 1948) up to the *Oslo Recommendations Regarding the Linguistic Rights of National Minorities* (OSCE, 1998). These international and national laws, conventions, agreements and regulations create the legal basis for the formation of multicultural integration practices in Russia. However, the average Muscovites' preferences are rather more prone to the assimilation model. That is why the experts emphasized the importance of the mechanisms and structures that should provide for the improved information on migration, on the procedures for naturalization, employment, accommodation, access to health services, etc. The creation of better living and working conditions for migrants also requires the public hearing of the migrants themselves. Migrants should be given the opportunity to articulate their interests and difficulties in Moscow.

Two principal strategies of the respondents may serve as orientations for the future administrative decisions. The first strategy is to stay in Moscow. This strategy is being pursued by 72.5% of the migrants from Armenia and 59.5% of the migrants from Georgia. Thus, nearly three quarters of the ArGeMi respondents from Armenia and more than the half of those from Georgia basically consider Moscow as the place of their permanent residence. Such migrants are ready for integration into the Moscow community and undertake measures to this end on their own initiatives. It is obvious that they may augment the social potential of the city and that will not require high expenses or efforts on the part of the city. The second strategy is to earn as much as possible and to return to the country of origin.

Migrant associations are important factors of integration. However, there is also the opinion that such labor or cultural associations rather prevent the integration into the host communities. Moreover, there are cases when immigrant communities establish informal networks with the aim to establish and retain control over lucrative commercial spheres like for example markets. The leaders of ethnic communities often have the financial resources and organizational abilities to achieve this. The activities of such networks aim at the division of spheres of influence among immigrant communities (see Tyuryukanova 2006). Nevertheless, there are also numerous positive examples of the migrants' associations influence on their integration into the recipient communities. Such associations support the newcomers to better orientate themselves, to gain access to health services and accommodation. As seen from another point of view, migrant associations preserve the cultural autonomy of ethnic groups in terms of preserving their native language, education, and national culture. Migrant associations are represented in the *Assembly of Russia's*. The ethnic associations interact with state bodies and local authorities and support multicultural social practices.

The mass media play a particularly important role in the migrants' integration into the host community. The interviewed experts noted that the Russian media tend to present migrants as generally uneducated, uncultured 'guest workers', or as criminals. Migrants from Armenia are frequently presented as hard-working and rich people, mostly engaged in commerce. But migrants from Georgia are usually presented as aggressive and involved in crime. The negative stereotypes cause a cautious perception of migrants among a considerable part of the majority population, and subsequently xenophobia and hostility towards them. However, the interviewed experts were not always unanimous in their assessment of the situation. They also shared the opinion that the attitudes of different strata toward the migrant groups were different, and that in general the attitude toward migrants from Armenia and Georgia was tolerant, normal and calm. One expert mentioned that migrants from Georgia were more comprehensible to Muscovites and their character was more akin to Muscovites. But practically all interviewees admitted that the media campaigns had formed a generally negative image of South Caucasians, in particular with regard to migrants from Georgia, who are feared as criminals.

The monitoring of events performed within the framework of the ArGeMi study during the period from February through April 2009 and from February through April 2010 focused on this media impact. The monitoring provided for the tracking of developments connected with cross-boundary migration and their coverage in Moscow. The primary information of the monitoring was obtained from publications in newspapers, documents of governmental agencies, economic, political and cultural organizations as well as from publications of research centers, NGOs and international news agencies. Of the 30 cases recorded in 2009 15 were devoted to problems of migrants from Georgia, eleven cases were devoted to general issues of migration to Moscow and only four cases were devoted to issues of migrants from Armenia together with information about the cooperation of Russia and Armenia in this field. The migration related publications on Georgia could be divided into three groups: The first one was connected with the problem of visas for refugees from Abkhazia and with the increasing number of Georgian citizens who applied for asylum and also with the termination of diplomatic relations between Russia and Georgia. The second group was related to the formation of a positive image of Georgia. The third group was focused on the criminal activities of migrants from Georgia.

The media published a petition of the *Union of Georgians in Russia* presented to the Ministry of Foreign Affairs and the *Federal Migration Service* of the Russian Federation. The *Union* requested visas for refugees who found themselves in an illegal situation after the termination of the diplomatic relations between Russia and Georgia, for the restoration of direct interstate transportation for the sake of the development of mutual business and cultural ties. The *Union* also requested to restore the economic links between Moscow and Tbilisi by restoring the direct flights between the two countries, by opening of a Russian trade representative office in Georgia, by re-opening the Russian markets for Georgian wines, mineral waters and agricultural products. The appeal of the *Union of Georgians in Russia* was also directed to Presidents D. Medvedev and M. Saakashvili.

Information about the *Week of Georgian Culture* and the *Visit to a Georgian Family* held in the Moscow *House of Nationalities* was also published. The *Extraordinary Congress of Georgians Residing in Russia* found a positive response in the media. Information about the representative character of the participants, the creative atmosphere of the *Congress*, the participation of leaders of the Russian Federation as well as the greeting message of the President of the Russian Federation to the participants of the *Congress* were widely covered by the media. However, the public statement by M. Saakashvili, the President of Georgia, that the principal export of Georgia to Russia was not the wine, but crowned thieves, mafia bosses, criminals and similar elements was followed by a great number of publications about alleged numerous Georgian criminal gangs in Moscow. Figures were adduced that half of the 1,200 so called crowned thieves in Russia were natives of Georgia. The Ministry of Interior of the Russian Federation even confirmed that Georgians predominate among the thieves residing in Russia.

Issues concerning the unsatisfactory conditions of migrants at their new place of residence and violations of migration laws resulting from this dissatisfaction, and of crimes committed by migrants were raised most frequently in media materials devoted to general issues of migration in Moscow. It was reported that in Moscow alone in 2008 foreigners committed 30 percent of all crimes. These crimes were committed mainly by migrants from the former Soviet republics. In 2009, the crime rate rose by 12%. The media often raised questions concerning the growing inflow of irregular migrants to Russia during the period under scrutiny. It was noted that illegal employment was the most common of the three components of irregular migration to Russia (illegal entry and illegal residence being the two other). The problems of irregular migration were considered in close connection with the need to improve the efficiency of migration laws.

Some media coverage was devoted to the reduction of entry quota for migrants from the CIS member states. The authors of these publications used to recommend the reduction of the number of those migrants in the Russian Federation who do not have employment or means of subsistence. The media emphasized the increase of the vulnerability of migrants since they perform the major part of their activities through informal ties (relatives and friends), as well as through shadow agencies in the sphere of migration organization and employment. Migrants do not have a developed legal awareness and prefer not to insist on their rights or to do that through the same informal (or simply criminal) agents. The media also covered the efforts of the *Federal Migration Service* to control and manage migration by attaining a maximum benefit from migration and protecting the migrants whose needs and requirements should be taken into account under the conditions of the economic crisis. Several publications were devoted to the integration of migrant children into the educational process on equal terms as the Moscow born children. The experience of such work conducted by Russian social and educational institutions in Moscow as well as by Armenian and Georgian communities were summarized.

Furthermore, the media focused on the cash flows from Russia to Armenia and Georgia. Due to the crisis, a considerable decline of the official cash flows from Russia to these states was noted. From February through March of 2009 $ 164.2 million were transferred to Georgian commercial banks by using the electronic system of money transfers. This amount was by $ 40 million (or 19.6%) less than the remittances of January through March 2008. The migrants from Armenia, Azerbaijan and Georgia were reported to be leading in the average amount of transfers ($ 919, 825, 804 million respectively).

Two of the published materials on Armenia were devoted to issues of the economic relations between Russia and Armenia, to Russian investments in the construction of a new unit of the Gazprom owned nuclear power plant in Armenia, to the geological exploration and development of uranium minefields and the construction of a railroad linking Iran with Armenia. Two further publications were devoted to crimes against migrants from Armenia.

In 2010 the media reported about events connected with Georgia's withdrawal from the *Commonwealth of Independent States*. It was noted that the Georgian leadership substantiated this decision by the formal character of the Inter-parliamentary Assembly. The journalists noted, however, that Georgia intended to remain a party in 75 treaties and agreements concluded with the Assembly.

The intentionally false message of the Georgian TV channel *Imedi* about an invasion by Russian forces into Georgia and the alleged murder of M. Saakashvili caused an immense resonance in Russian print media and TV channels. It was pointed out that the Georgian TV channel was controlled by the Georgian authorities. Therefore, the broadcast was understood as a deliberate political provocation aimed at further aggravating the relations with Russia. The media published the opinion expressed by members of Georgia's political opposition who accused M. Saakashvili of his desire to unleash a new war against Russia. The media also broadcasted accusations for cooperation of Georgian authorities with terrorists as well as for transfers of drugs from Afghanistan to Russia via Georgia. The arrest of three Russian women who used to be citizens of Georgia but later became nationals of Russia and wanted to visit their dying mother in the city of Gori was widely covered by the Russian media as well. Some articles dealt with 'undiplomatic' characterizations given by the President of Georgia to the Russian leadership and with accusations brought by the RF special services against the leadership of Georgia.

In spite of this wave of mutual accusations the most important reported events during the scrutinized period were related to Russian and Georgian attempts at restoring the peaceful relations between the two countries. The point was typically raised that this development would have a beneficial impact on the migration exchange between Russia and Georgia and on the improvement of the situation of the Georgian migrants in Moscow. Much attention was focused on the steps made toward this direction by both countries, their political parties and organizations, specific leaders. A special events in this respect were the opening of the Georgian-Russian border which was achieved through the mediation of Armenia, the re-opening of the land communication of Russia with Georgia, and the permission of Easter flights from Georgia to Moscow.

The media paid particular attention to four events which were related to Armenia. It was noted that in 2009 82 thousand Armenian citizens left for Russia. Following the initiative of Russia, the *Council of Heads of Migration Bodies* of the member states of the CIS developed the legal document *General principles and mechanisms of labor migrants organized attraction for performance of labor activities in the member states of the CIS*. On this basis, an inter-agency agreement was signed between the *Federal Migration Service* and the police of the Republic of Armenia on cooperation in the struggle against irregular migration. The declaration on cooperation, signed by the Russian and Armenian Churches was also reported and discussed. The declaration presented a vision of the future relations between two national churches aimed at the protection of Christian values at international, national and regional levels, and at countermeasures against pseudo-religious trends. The exchange of experience concerning church services for the military, the youth, prisoners, and for the sick and aged was also foreseen. There were reports about intentions to open a representative office of the Russian Orthodox Church in Yerevan and of the Armenian Apostolic Church in Moscow. The reports in commemoration of the victims of the Sumgait pogrom (26-29 February 1988) were an expression of grief and respect. Media news informed that representatives of political, cultural and religious circles of Armenia and Russia took part in the event. An attempt upon the life of a businessman of Armenian origin was also reported in the publications under scrutiny. It was pointed out that the assault was obviously an attempted contract murder.

Documents of the *Federal Migration Service*, the Government of Moscow City and media reports characterized the activities of the FMS in Moscow and in Russia as aimed at the control of migration and the protection of the Russian national interest under the conditions of economic crisis.

7 Summary and Conclusions

The migration processes in the megalopolis of Moscow reflect the progress of social life toward respect of the human rights of free movement and correspond to the needs of the urban economy. Being an important source for preservation and development of the demographic potential of Moscow, migration to Moscow manifests all main types of migration (resettlement, seasonal, push-pull migration, episodic) and categories of migration (economic, forced, familial, educational etc.). The special character of the migration to Moscow is conditioned by geographic selectivity, by the growth of the number of the labor, temporary and irregular migrants, by the settlement of migrants who work in Moscow beyond its administrative frontiers in the Moscow region, as well as by the Moscow authorities' predisposition to restrictive migration policies.

The intensification of the migration exchange and the changes in migration types after 2004 should be referred to economic incentives as the predominant motif of the current migration from Armenia and Georgia to Moscow. Another specific trait of the contemporary migration flows from both countries to Moscow is the decline of forced migration and the increase of voluntary, unorganized, temporary, recurrent and educational migration. The main causes of migration movements from Armenia and Georgia are the wide opportunities for work and business in Moscow, the high level of wages there which allows remittances for support of friends or relatives, and the good command of the Russian language by the migrants. The findings of the ArGeMi study revealed that the migrants' main strategy was to stay in Moscow for an extended period and most probably permanently. Migrants from Georgia are more often inclined to temporary stays in Moscow, while migrants from Armenia prefer permanent residence. The ArGeMi study confirms the hypotheses that in the future there will be no perceptible change in the migration exchanges of the Russian Federation with both countries under investigation. Neither can massive return of migrants to Armenia and Georgia be expected. The character of migration from Armenia and Georgia to the Russian Federation largely depends on the interstate relations of both countries with Russia.

In most cases return migration from Moscow to Armenia and Georgia bears a temporary and voluntary character. The migrants would like to see their relatives and families or return temporarily for other personal reasons. In times of economic crisis other reasons were mentioned, too, like the lack, completion or loss of employment, problems with the law of the Russian Federation and the high living expenses in Moscow. Xenophobia and corruption were also mentioned as return reasons. Some migrants from Armenia and Georgia in Moscow intend to depart for third countries. In order of preferences, the USA, Germany, Israel, Canada, Greece, France, Australia, Bulgaria, Czech Republic, Lebanon and Iran were mentioned as country of favorable destination.

The study identified also certain differences between the migrants from the two South Caucasian states. The average profile of the migrant from Armenia was established to be the following: A male or female up to 35 year old, unmarried, and an ethnic Armenian. The average profile of a migrant from Georgia was a male up to 35 years, unmarried, ethnic Russian, Georgian or Armenian. The marriage status of the female migrants was somewhat different from that of the males. There were less unmarried women and more divorced and widowed female migrants among the ArGeMi respondents.

The judgments of the experts confirmed the decrease of the traditional gender disparity in the migration flows. A trend of feminization in the migration from Armenia and Georgia to Moscow was noted. According to the experts, in future the ratio of women among the migrants from both countries will further rise. The overwhelming majority of the migrants had an urban background. They were basically well prepared to the life in a large city. Due to their young age many migrants did not have to care about households or about children in their country of origin. Some of them did not have a place of residence in their country of origin prior to their arrival in Moscow.

Moscow's labor market shows patterns of employment that are typical for a recipient developed country. As a rule migrants are employed in trade, construction and transportation. These are work places which are not attractive to the permanent residents of Moscow. A rather big share of the employed ArGeMi respondents (19% of the working migrants from Armenia and 15.3% of the working migrants from Georgia) were employed in "other" (unspecified) spheres. According to the experts, migrants from Armenia are employed in the sphere of science, culture, accommodation and utilities infrastructure or are engaged in small and medium businesses, consumer services (footwear and watch repair). The migrants from Georgia would typically work as private taxi-drivers or in other transportation spheres, in the restaurant and gambling business, in cultural, entertainment and leisure activities. In comparison with migrants from Central Asia, the experts noted the high level of education and diversified employment among migrants from Armenia and Georgia. 7.7% of the respondents from Armenia and 16.3% of those from Georgia had managed to establish their own firms.

The migrants from Armenia and Georgia try to select these sectors of the Moscow economy for their employment which at least partially correspond to their previous work experience. However, the choice is also determined by the requirements of Moscow's labor market. The findings of the ArGeMi study confirm the significance of unofficial channels and informal relations for bridging often diverging factors of the decision for employment. As seen from another point of view, the same findings confirm the inefficiency of the formalized system of employment provision for migrants. Such a system of migrant employment support is practically missing in Armenia and in Georgia. Our study also bears witness to an obvious insufficiency of information on issues related to immigration and the legalization of migrants in the labor market of Moscow. To a considerable extent the choice of a branch for employment or business was determined by the migrants' own initiative. The efficiency of individual efforts in the search for employment illustrates, among others, the high potential of the metropolitan labor market.

The employment of more than every fourth ArGeMi respondent from Armenia and Georgia did not at all correspond to the type and the level of their education. The partial match of the educational level and the employment is observed in approximately 40% of all cases. The employment of women more often did not correspond to the type and level of and the previous professional qualifications of the female migrants from Georgia. The explanations might be the higher level of the female education and the somewhat narrower range of employment opportunities for female migrants in the Moscow labor market. Despite this obvious "brain waste", the overwhelming majority of interviewed respondents (9 of 10) were 'partly' or 'entirely' satisfied with their employment in Moscow. The 'entirely satisfied' made out 33% of the respondents from Armenia and 25.2% of those from Georgia. The employment provides the respondents and their families with relatively high incomes. They are similar to the average per capita incomes of Muscovites.

While staying in Moscow, migrants from Armenia and Georgia retain close ties with relatives and friends in their countries of origin. About a half of the respondents transferred money to their families, relatives and friends in Armenia and Georgia. These remittances play an important role in coping with the economic hardships that large parts of the population face in both countries. Every third respondent from Armenia and every fifth from Georgia used to maintain contacts with compatriots who had settled in the capital of Russia and used to visit an ethnic club or organization at least once a month. At the same time, the ArGeMi respondents were mostly involved in the Russian information context. As a rule, the respondents used to read Russian printed media and watched mainly Russian TV. Information from the Armenian or Georgian media was consumed by every fourth migrant once a week.

The social contacts of a migrant may serve as an indicator of his or her integration into the local social milieu. The ethnic Russians predominate among the friends of the respondents. Compatriots from Armenia and Georgia who had settled in Moscow made out the next group of frequently mentioned friends in Moscow. In cases of crisis or emergency, the respondents preferred to apply for help first of all to their local acquaintances, then to compatriots and on the third place to the local authorities. A surprisingly insignificant share of respondents would seek help from their national churches, national organizations or national embassies. However, the majority of the migrants prefer to have spouses belonging to their own ethnic group.

The attitude of the average people in Moscow to them had been appraised by the respondents in a remarkably positive way. 53.5% of the respondents from Armenia and 45.5% of those from Georgia defined them as “friendly”. Only 2.5% and 7.5% respectively qualified this attitude as “unfriendly”. This assessment is rather important since during their stay in Moscow 32% of the migrants from Armenia and 38% of those from Georgia had been verbally offended or even physically attacked (8.5% and 5% respectively). The explanations why a minority of respondents (11% of the migrants from Armenia and 26% of those from Georgia) felt insecure in Moscow were ranged as follows: Muscovites dislike Georgians, foreigners in general and labor migrants in particular. When assessing the attitude of the authorities towards them, nearly every third respondent from Armenia and every fifth migrant from Georgia reported a friendly attitude, while 12% of the migrants from Armenia and 17% of those from Georgia defined this attitude as unfriendly. The prevailing experience of the authorities' attitude is indifference (55.5% and 57.5% of the respondents respectively).

According to the overwhelming majority of experts, migrants from Armenia and Georgia do not possess and exercise their human and social rights and experience often troubles with the police or officers from the *Federal Migration Service*. Systematic informational and legal support to migrants is very much needed together with comprehensive programs for legalization, provision of social and labor guarantees for migrants, provision of education for their children. The FMS offices and officers should be more active and efficient in their cooperation with the ethnic communities and the organizations of migrants for the prevention of illegal residences and other irregularities accompanying the stay of migrants in Moscow.

Over the past two decades under scrutiny the migration policy of the Russian Federation underwent substantial transformations. In the 1990s the migration was practically not regulated by the state. In the 2000s the state pursued a policy of severe migration restriction. This restrictive policy was partly reversed by liberalization in 2007, but in 2009 a new strengthening of the restrictive policy came about due to the global economic crisis. The experts believe that this restrictive migration policy could not be long-termed. They interpreted it mostly as a lack of a clearly defined strategy for managing cross-boundary migration since the practiced migration policy was unanimously described as far from protecting national interests. The experts stressed the point that the leadership of the Russian Federation had recognized the necessity of a more liberal policy, particularly in the sphere of labor migration. Some measures aimed at migrant integration were already undertaken. However, these efforts were evaluated by the experts as insufficient, particularly concerning the integration of the migrants in the social context of the Moscow community. Lack of clarity was identified in the decisions about who and in which way should be supported for integration as well as which administrative body should be basically responsible for this policy.

The assessment of the experts concerning the policy of the Government of the Russian Federation was ambiguous. It was frequently noted that the government "encourages and impedes migration in the same time". With regard to the migration policy of the Moscow City authorities the statement "impedes" prevailed. The experts stressed the importance of immigration as a resource for the development of the city. It was clearly formulated that the metropolitan megalopolis could not function without labor migrants. The city needs skilled and unskilled migrants, but the demand for unskilled migrants is particularly strong. They perform works that the resident Muscovites refuse. However, the experts also emphasized that immigration is always fraught with risks for the migrants themselves. The experts noted the low level of tolerance in Moscow, the corruption in the law enforcement bodies and the widespread negative attitude toward migrants from the Caucasus. The topic concerning the risks for the recipient country was elaborated in details as well. The universal issues related to the impact of immigration on the rise of crime, overpopulation and interethnic tensions were discussed with a view to the local situation in Moscow. According to the experts, semi-legal and irregular migration generates the greatest risks while regular migration is safe. The main way to reduce risks is an effective control of migration processes, combined with a reasonable integration policy.

The Government of Moscow City had made certain steps into this direction. The organization of arrivals on the "one window" principle and the establishment of simplified type hostels where migrants can register and reside were evaluated as promising ideas which were incorporated into the Moscow migration program for 2008-2010. However, the program did not foresee the whole range of measures which are needed for the full integration of the migrants into the Moscow City community.

The legal basis of the migration policy was evaluated as still too general and not determined enough. Its major flaws were seen in the insufficient protection of migrants' rights since the legal regulations rarely stipulate precise mechanisms and procedures of law enforcement. Another important flaw in the law enforcement was seen in the lack of clear mechanisms for coordination among the institutions responsible for the law enforcement.

The role of the media for an efficient and civilized management of migration processes was repeatedly put into the center of discussions. It was noted on various occasions that the media shape the public opinion about migration and migrants. This point was stressed since the mass media in Moscow were often blamed for generally presenting migrants as uneducated, uncultivated guest workers and as criminals. This stereotyping has consequences for the perception of the migrants from Armenia and Georgia in Moscow. The experts repeated that migrants from Armenia were usually presented as diligent and rich people who are mostly engaged in commerce. Migrants from Georgia were typically presented as aggressive and rich people who are often involved in crime. These stereotypes were identified as dominating the perception of migrants from Armenia and Georgia. The stereotypes cause and reproduce xenophobia and hostility against migrants in general and against migrants from Armenia and Georgia in particular.

Given the unavoidable increase of the demand for replenishment of the population and of labor resources in Moscow, ***the principal direction of the improvement of the migration policy*** should be the improvement of the legal basis and the mechanisms of law enforcement. It was repeatedly stressed by the experts that the whole package of laws related to migration should be brought into compliance with present day realities and the international norms for regulation of trans-boundary migration. The experts insisted that it is mandatory to adopt the *Convention on Rights of Workers and Members of Their Families in the CIS Countries*, to update the bilateral *Treaty between Russia and Armenia* and to conclude a similar agreement with Georgia. The experts also recommended the elaboration of legal acts that regulate the transfer of pensions, the insurance, indemnities and other issues. The regulation of these issues by taking into account the differences in the national laws of Armenia and Georgia could not be postponed any more. In addition, standardized data collection concerning cross-boundary migration should be established in order to make rational management of the migration processes possible.

In more specific terms, it was proposed to develop a strategy for the management of migration processes in the capital city of the Russian Federation and to secure effective mechanisms for the implementation of this strategy. More precisely, it should include:

- Improved interaction with the Armenian and Georgian authorities in the regulation of migration flows in order to attract to Moscow those migrants that the city really needs and to provide them with the full package of social guarantees;
- Improvement of the efficiency of migrants' employment;
- Development of the integration component of the migration policy in order to provide for equal opportunities and effectiveness of migrant incorporation into the Moscow community;
- Diversification of the economic, political and cultural measures contributing to the integration of migrants into the local community.

Following the above principles of local management of the trans-boundary migration, it was required to specify the medium-term migration program of the Moscow City Government. According to the experts it should be aimed at the improvement of registration procedures, at naturalization, overcoming corruption, enhancement of employers' responsibility for the provision of migrants with the necessary conditions of accommodation and work. The experts particularly stressed the need for selective programs attracting skilled labor. A special program aimed at strengthening the cohesion of the Moscow community was also seen as mandatory. The program should provide for the efficient adaptation and integration of migrants. It should also contribute to the development of a tolerant attitude to people having different ethnic and cultural backgrounds.

References

ČISLENNOST' i migratsiya naseleniya Rossiyskoy Federatsii v 1998: Statističeskiy bulleten' [Number and migration of the population of the Russian Federation in 1998; Statistical Bulletin] (1999): Moscow: Goskomstat

DEMOGRAFIČESKAYA SITUATSIYA *v Rossiyskoy Federatsii v 2008 godu* [Demographic situation in the Russian Federation in 2008] (2009): Moscow: Goskomstat
http://www.gks.ru/free_doc/2009/demo/demo-docl08.htm.
Accessed on 4 February 2010.

DEMOGRAFIČESKIY EŽEGODNIK [Demographic yearbook] 2001-2009 (2010): Moscow: Goskomstat
http://www.gks.ru/wps/P_A_S5/Documents/jsp/Detail_defoult.jsp?category=1112178611292&elementid=11374209312. Accessed on 02 March 2010

DOKLAD o razvitii čelovečeskogo potentsiala v Rossiyskoy Federatsii, 2008: Rossiya pered litsom demgrafičeskikh vyzovov [Report on human potential development in the Russian Federation in 2008: Russian in the face of demographic challenges] (2009) Moscow
http://demoscope.ru/weekly/knigi/undp2008rus/undp2008.html

DROBIŽEVA, Leokadiya, Yuriy ARUTYUNYAN and Igor' KUZNETSOV (2007): 'Vykhodtsy iz Zakavkaz'ya v Moskve' [Natives of Trans-Caucasus in Moscow]. *Demoskop Weekly*, No. 271, 1-21 Yanvarya http://demoscope.ru/weekly/2007/0271/tema03.php

Dvustoronnye soglašeniya Rossiyskoy Federatsii [Bilateral Agreements of the Russian Federation] (2010) Moscow http://www.sibupk.nsk.su/New/06/Migration/1_2_dvustoronn_akty.htm. Last access on 12 February 2010

DYNAMICS of the Social sphere of Russia: Realities and Prospects (2009): Moscow: The Russian State Social University

Federal'naya služba gosudarstvennoy statistiki: Informatsiya o sotsial'no-ėkonomičeskom položennii v Rossii [The Federal State Statistical Service: Information on social-economic situation in Russia]. Section VII: Demography. http://www.gks.ru/bgd/free/b09_00

FILIPPOV, V. (2003): *Dinamika etničeskogo i konfessional'nogo sostava naseleniya Moskvy po dannym predydushchikh perepisey* [Dynamics of ethnic and confessional structure of Moscow population according to census data]. In: Tiškov, Valery (Ed.): Na puti k perepisi [On the path to the census]. Moscow: Aviaizdat, pp. 277-313

Gornyie strany: rasselenie, etnodemografičeskie i geopolitičeckie processy, geoinformatsionnyi monitoring [Mountain countries: settlement, ethnodemographic processes: geo-information monitoring] (2005): Moscow-Stavropol: Stavropol State University Press

IGNATOVA, O. (2010): 'Mimo kvot' [Missing the quote]. *Rossiyskaya gazeta*, № 5115 (36) 19 February. http://www.rg.ru/2010/02/19/migrant.html Accessed on 28 August 2010

IVAKHNYUK, IRINA (2009): *Russian Migration Policy and Its Impact on Human Development of Population*. Moscow: Moscow Lomonosov State University. MPRA Paper No. 19196. http://mpra.ub.uni-muenchen.de/19196/

KARPENKO, O. M., M.D. BERSHADSKAYA and Yu. A. VOZNESENSKAYA (2008): 'Pokazateli urovnya obrazovaniya naseleniya v stranakh mira: analiz dannykh mezhdunarodnoy statistki' [Indices of level of popular education in countries of the world: Analysis of international statistical data]. *Sociology of Education,* No. 6, pp. 4-20

KOZAKOV, A. (2005): 'Kto učastvuet v detorodnom processe?' [Who does take part in the birthgiving process?]. *Kvartirnyi ryad*, 1 September

'Krizisnaya migratsiya: kto pokinul Moskvu v 2009' [The crisis migration: who did leave Moscow in 2009] (2010). 7 January http://www.newsmsk.com./article/07Jan2010/migr_2009.html.

LEBEDEV, A. (2006): 'Moskva kak stolitsa migrantov?' [Is Moscow the capital of migrants?] *Polemika i diskussii* [Polemics and Discussions] 31 October http://www.polemics.ru/articles/?articleID=9292&hideText=0&itemPage=1. Accessed on 25 August 2010

Moskovskiy meždunarodnyi portal: Demografičeskaya situatsiya [Moscow international portal: Demographic situation] (2009) http://infrastructure.moscow.ru/ru/economy_business/megapolos/social_portret/demog...

Moscow City Program for the Years from 2008 to 2010 (2007): Moscow: The Government of Moscow City Resolution No. 711 of 21 August

MOSKVA i migratsiya [Moscow and migration] (2009) http:www.polit.ru/research/2009/10/15/demoscope389.html

MUKOMEL, Vladimir I.: Demografičeskie posledstviya etničeskikh i regional'nykh kofliktov v SNG [Demographic consequences of ethnic and regional conflicts in the CIS] (1999). *Sociologičeskie issledovaniya,* No. 5, pp. 66-71

On Approval of Rules of Registration and Registration Withdrawal of the Russian Federation Citizens at Places of their Staying and Residence in Moscow City (2004) Moscow: Resolution No. 189 of the Government of Moscow City, 6 April

On Approval of the Quota for Issuance of Work Permits for Foreign Citizens in 2009 (2008) Moscow: Resolution No. 835 of the Government of the Russian Federation, 7 November. http://www.visa-workpermit.ru/128.0html

On the citizenship of the Russian Federation (2002) Moscow: Federal Law No. 62 of the Russian Federation, 31 May

On Distribution of the Quota for Issuance of Work Permits for Foreign Citizens Approved by the Russian Federation Government for 2009 (2008) Moscow: Order No. 777 of the Ministry of Social Development and Public Health of the Russian Federation, 26 December

On the Legal Status of Foreign Citizens in the Russian Federation (2002) Moscow: Federal Law *No.* 115, 25 July

On Migration Registration of Foreign Citizens and Stateless Persons in the Russian Federation (2006) Moscow: Federal Law of the Russian Federation No. 110, 17 July

On the Procedure of Registration of Foreign Citizens Employed at Urban Markets, Fairs, Trade Complexes and Centers and Trade Houses Located in the Territory of Moscow City (2001) Moscow: Decree No. 447 of the Mayor of Moscow City, 7 May

On the Procedure of Registration and Execution of Foreign Citizens Employed in the Service Sphere (2000) Moscow: Decree No. 1090 of the Mayor of Moscow City, 13 October

O razvitii čelovečeskogo potentsiala v Rossiyskoy Federatsii v 2008 godu: Rossiya pered litsom demografičeskikh vyzovov ['On the Development of the Human Potential in the Russian Federation in 2008: Russia Facing the Demographic Challenges] (2009): Moscow: Government of the Russian Federation. http://demoscope.ru/weekly/knigi/undp2009rus/undp2008.html. Access on 17 February 2010

OSADCHAYA, GALINA I. (Ed.) (2009): Dinamika sotsial'noy sfery rossiyskogo obščstva [Dynamics of the Social Sphere of the Russian Society]. Moscow: The Russian State Social University Press

OSADCHAYA, GALINA I. (Ed.) (2008): Dinamika sotsial'noy sfery rossiyskogo obščestva [Dynamics of the Social Sphere of the Russian Society]. Moscow: The Russian State Social University Press

OSADCHAYA, GALINA I. (Ed.) (2007): Dinamika sotsial'noi sfery rossiyskogo obščestva [Dynamics of the Social Sphere of the Russian Society]. Moscow: The Russian State Social University Press

Osnovnye pokazateli denežnykh dokhodov naseleniya v real'nom vyraženii [Main Indices of Money Incomes of the Population in Real Terms] (2009) Moscow: Territorial'nyi organ Federal'noy služby gosudarstvennoy statistiki po gorodu Moskve [The Federal State Statistical Service Territorial Body for Moscow City] http://moscow.gks.ru/digital/region12/DocLib/pokazdoh.htm. Access 30 April 2010

PALEEV, Anton (2009): 'Avral'noy migratsii nam ne nado' [We Don't Need an Emergency Migration]. *Moskovskiy komsomolets*, 15 September

POLYAN, P. M. and T. SELIVANOVA (2005): 'Gorodskie aglomeratsii Rossii i novyie tendentsii evolutsii ikh seti (1989-2002)' [Urban Agglomerations in Russia and New Trends of Their Evolution]. In: Gornyie strany: rasselenie, etnodemografičeskie i geopolitičeskie protsessy, geoinformatsionnyi monitoring [Mountainous countries: settlement, ethno-demographic processes, geoinformational monitoring: Materials of the international conference], Stavropol-Dombai: Stavropol State University Press, pp. 290-296

ROMODANOVSKY, Konstantin (2008): Vystuplenie direktora FMS Rossii [Presentation made by the FMS Director], 13 February. Moscow: FMS www.fms.gov.ru/press/publications/news_detail.php?ID=9792

ROMODANOVSKIY, Konstantin (2010): *FMS v usloviyakh krizisa sumela zaščitit' rossiyan ot nenužnoy konkurentsii s migrantami na rynke truda* [The FMS managed to protect Russians from unnecessary competition with migrants in the labor market]. Moscow: FMS http://www.migrant.ru/news.php?id=584

Rossiyskoe ėkonomičeskoe čudo: sdelaem sami [The Russian Economic Miracle: We'll Do It on Our Own] (2007): Moscow: Center for Macro-economic Analysis and Short-term Forecasting

RYBAKOVSKIY, Leonid L. (Ed.) (2006): *Demografičeskaya situatsiya v Moskve i tendentsiya ee razvitiya* [Demographic situation in Moscow and the trend of its development]. Moscow: The Center for Social Studies

Rynok truda: Statistika [Labor Market. Statistics] (2009): Moscow: Departament truda i zanyatosti [Department of Labor and Employment] http://www.labor.ru/?id=1012 . Accessed on 12 February 2010

SMIDOVICH, S. G. (2005): 'Problemy regulirovaniya migratsii v Moskve' [Problems of Regulation of Migration in Moscow]. In: *Migratsionnyi process:*

prošloe, nastoyaščee, buduščee [Migration Process: Past, Present, Future]. Moscow: The Russian State Social University, pp. 57-68

Territorial'nyi organ Federal'noy služby gosudarstvennoy statistiki po gorodu Moskve [The Territorial Body of the Federal State Statistical Service for Moscow City]. Available at: http:Moscow.gks.ru/default.aspx

'3000 učaščikhsya izučayut russkiy yazyk kak inostrannyi' [3000 Students Learn Russian as a Foreign Language] (2006): *Mosinform*, 26 June www.educom.ru/department/news/news_detail.php?ID=4329

TYURYUKANOVA, ELENA V. (2006): 'Migranty na neformal'nom rynke truda v Moskve' [Migrants at the Informal Labor Market of Moscow]. *Nauchnyi žurnal*. The Russian State Humanitarian University Press, Vol. 8, pp. 51-59

TYURYUKANOVA, ELENA V. (2009) Novaya migratsionnaya politika v oblasti trudovoy migratsii: pervye itogi. [New Migration Policy in the Sphere of Labor Migration: The First Results]. http://www.fms.gov.ru/about/science/science_session/forth/tur.pdf.

VOZMITEL, A. A. and Galina I. OSADCHAYA (Eds.)) (2009): Obraz žizni v sovetskoy i postsovetskoy Rossii: dinamika izmenenii [Way of Life in the Soviet and post-Soviet Russia: Dynamics of Changes]. Moscow: The Russian State Social University

Vserossiyskiy tsentr Izučeniya obščestvennogo mneniya [All-Russian Center for Public Opinion Research] Press-release 1034 (29.08.2008) and 1038 (04.09.2008) http://wciom.ru/novosti/press-vypuski/browse/35.html?L%5B0%5D=0%26cHash%3D6ce1572e1&cHash=f0d63da9fe. Accessed on 28 August 2010

ZAYONČKOVSKAYA, Žanna and Nikita MKRTČYAN (2009): 'Moskva – dinamično rastuščiy megapolis' [Moscow - a Dynamically Growing Megapolis]. *Demoskop Weekly*, 15 October, http://www.polit.ru/research/2009/10/15/demoscope389.html

Žebit, Mariya (2009): 'Priezžayte, tut vas ždut!' [You are welcome, here you are expected] *Rossiyskaya Gazeta* No. 4988 (164), 3 September. http://www.rg.ru/2009/09/03/migranty.html. Last access on 28 August 2010

ZUBAREVIČ, Natalia V. (2007): 'Aglomaratsionnyi effect ili administrativnyi ugar?' [An agglomeration effect or administrative frenzy?], *Rossiyskoe ėkspertnoe obozrenie*, No. 4-5

Tessa Savvidis

Comparing Out-Migration from Armenia and Georgia

In the ArGeMi survey, migration was defined as any cross-border movement lasting more than a month, regardless of the motives of such movements. Subsequently, a migrant could be a person who left his or her country of origin for personal or familial, educational, professional, or political reasons. He could be a student, a refugee, an employee in search of labor and employment, a businessmen or a spouse joining his or her partner abroad. The volume of each migratory sub-category obviously depends on political, socio-economical and legal circumstances and varies in diachronic comparison.

The 77 questions of the ArGeMi questionnaire for six of the eight cohorts of migrant interviewees had been broken into seven items (only five for the two samples of potential migrants, respectively). These items were:

- Passport data of the respondent
- Education and professional status
- Social situation of respondent and his/her household
- Personal migration experience during previous travels of respondent
- Evaluation of personal experience in respondent's usual country of foreign destinations
- General evaluation of migration and emigration
- Respondent's future migration plans.

Every migratory movement takes place in a given historical, legal, political, and cultural environment. Most of the respondents, who have been interviewed in the ArGeMi research project, move or have migrated in the post-Soviet space or the CIS migration system, respectively. After a brief historical introduction and analysis of this particular migration system, poverty and ethnic or territorial conflict constellations are analyzed as some of the specific main push factors that have caused massive flight in the first post-Soviet decade, especially in the South Caucasus. This is followed by a comparative analysis of some key results of the ArGeMi polls on various levels: a) from the three fields Armenia, Georgia, Moscow, b) the returnees and Moscow cohorts, c) the returnees from the post-Soviet migration space and from other destinations, d) between factual migrants (returnees and cohorts in Moscow) and potential migrants. In addition, the ArGeMi results will be related to earlier polls, or they are complemented by subsequent polls. In conclusion, we re-consider the traditional perception of Moscow as an 'extended home-land' by numerous migrants from the South Caucasus, scoring this perception against the increasingly xenophobic reality in this megalopolis.

1 Migration before 1991: From Restricted to State Promoted Mobility

In the Russian Empire, migration processes were of centrifugal colonial character (Ivakhnyuk 2009: 4), with voluntary and compulsory (re)settlement of populaces to the 'okrainy', or fringes, while the mobility of the rural majority was restricted until 1861 by the feudal *krepostnoe pravo* (literally, 'fortress law'); *de facto*, serfdom of the Russian peasantry existed since the 17th century, and *de iure* since 1723. Despite the immobility of the vast majority of Russia's population, the borderlands and newly acquired territories were strengthened by the replacement or the 'dilution' of indigenous populaces, as it happened in the predominantly Muslim Western and Central parts of the Northern Caucasus in the mid-19th century.

Throughout the Soviet period, migration inside the USSR was existent, but state-controlled, with up to two million persons annually being involved. In contrast to the primarily strategic and demographic aims of the pre-Soviet migration management the Soviet migration regime has been labor oriented. Its primary aim was the balance between labor excessive and labor deficit areas in the Soviet realm by re-distributing the population (Ivakhnyuk 2009: 4-5). Obligatory *raspredelenie* ("distribution") of recent university or high-school graduates – at least for two or three years - was one of the effective measures in order to secure a regular inflow of qualified professionals and a balanced qualification level, even in the most remote areas of Soviet Union. *Raspredelenie* and similar measures also caused considerate changes in the ethnic composition of local or regional populaces, because millions of highly skilled Slavic professionals from Russia, Ukraine and Belarus arrived in Central Asia and the Caucasus, while at the same time hundreds of thousands of professionals from those areas worked in other parts of the USSR. As a result of this multi-ethnization, 25 million ethnic Russians were living outside the RSFSR in the late 1980s (Narodonaselene 1994: 128).

While the mobility of highly skilled labor was state promoted, movements of unskilled labor were largely prevented by the notorious *propiska*, or registration system. Without that registration with the local registration office no worker could get a job or employment authorization. However, on the longer run this dual system undermined the emergence and development of the labor market, "due to the absence or underdevelopment of competition and free-market mechanisms in education and job hiring" (Abazov 2009: 8).

Despite such restrictions there existed an increasing informal or non-state sector of employment opportunities for seasonal or temporary migrants already in Soviet times, which can be regarded as a for-runner of the post-Soviet migration situation in the Commonwealth of Independent States. This sector comprised the construction sites, often in areas with harsh climate, and the retail sector, especially the *kolkhozniye bazary*, where surplus products of the agricultural cooperatives could be sold under near to free-market conditions. "In addition, Soviet authorities' notorious inability to provide a sufficient supply of agricultural products inspired many entrepreneurs from Central Asia, the Caucasus and to a lesser degree from Moldova and the Ukraine, to sell their products on the informal basis in the less-regulated *kolkhozniye bazary*" (Abazov 2009: 9).

Already in the 1980s these informal sectors of the Soviet labor market involved between 400,000 to 800,000 people per year, or 10 to 20 percent of the underemployed rural population in Central Asia and Caucasus (Abazov 2009: 9). Trade and other enterprises that emerged in the late Soviet period became the basis of more relevant private regional trade networks in post-Soviet times.

2 Post-Soviet Regional Migration - A System of Complementarity

In contrast to the state controlled and state organized population movements of the Soviet era, residents of the post-Soviet space enjoy a significantly higher degree of individual freedom, including the freedom of movement and travel. This freedom is highest where migrants move inside a common geographical, political, legal and economic space that facilitates their mobility. Such a space is the *Commonwealth of Independent States* (CIS) which in difference to the migration space of the European Union is not the result of economic and political integration, but in the first place a loose coalition that attempts to overcome the repercussions of political disintegration and failing central co-ordination. To some extent and on various levels the post-Soviet space has even undergone a process of political and economic re-integration since the formation of the CIS in 1991.[17] With regard to regional migration, the CIS can also be described as the jointly continued administration of the legacy of the USSR by its independent successor states, with a common, visa-free labor market as one of its most attractive elements. The founding member states of the CIS pledged to support the free flow of people, goods, and services within the Commonwealth. It is indicative for the significance of migration in the re-integration process that legislative acts on migration were among the first issued by the Commonwealth, with the framework for cooperation in migration being the *Almaty Declaration* (December 1991), the *Agreement on Establishing Consultative Council on Labor, Migration*

17 The member states of the CIS being: Armenia, Azerbaijan, Belarus, Georgia (until 18th August 2009), Kazakhstan, Kyrgyzstan, Moldova, the Russian Federation, Tajikistan, Turkmenistan, Ukraine and Uzbekistan. Of these, five countries (Belarus, Kazakhstan, Kyrgyzstan, Russia, Tajikistan) signed an Agreement on the creation of an Eurasian Economic Community (October 2000). At present Armenia, Moldova and Ukraine have the status of the observer under EAEC. In October 2005 Uzbekistan made the statement to join this organization. In September 2003 four countries - Belarus, Kazakhstan, Russia and Ukraine signed an Agreement on the Formation of CES (Common Economic Space). - http://www.cisstat.com/eng/cis.htm

and Social Protection (November 1992), the *CIS Inter-Government Treaty on Migration and Social Protection of Labor Migrants* (April 1994) and the *CIS Treaty on Cooperation against Illegal Migrants* (March 1998). In addition, between 1992 and 2000 most CIS members signed bilateral agreements that allowed a visa-free travel regime between the signatories. Belarus, Kazakhstan, Kyrgyzstan, Russia and Tajikistan confirmed the visa-free regime by a new multilateral treaty in 2005.

Russian migration scholar Irina Ivakhnyuk, based on her colleague Vladimir Mukomel, summarizes the main feature of the CIS migration space in the following way: "(...) migrants vote by their feet for a single migration space and a common labor market. In the 1.5 decades of post-Soviet development often complicated by contradictions of interests and lack of understanding, freedom of movement was likely to be the strongest link connecting the former Soviet republics" (Ivakhnyuk 2006: 8).

Migration flows inside the former Soviet space are mainly determined by visa regimes. The degree of the accessibility of countries of entry varies considerably, being lowest in Turkmenistan (TK), which has practically ruled out all other former Soviet states in the visa waiver. In its restrictiveness, Turkmenistan is followed by the three Baltic republics, which are now members of the EU. As a country of entry, Moldova (MD), on the other hand, offers free entry for all other previous Soviet republics, including the Baltic States. It is followed by Georgia (GE), which demands visas only from nationals of Turkmenistan, while Ukraine (UA) is the only large territorial state that stays visa-free for nationals of all previous Soviet republics (with the exception of the Baltic States):

Table 1: Visa regimes in former USSR republics

Country of entry	COUNTRY OF ORIGIN														
	RU	BY	MD	UA	AM	AZ	GE	KZ	KG	TJ	TK	UZ	LV	LT	EE
RU		--	--	--	--	--	V	--	--	--	V	--	V	V	V
BY	--		--	--	--	--	--	--	--	--	--	--	V	V	V
MD	--	--		--	--	--	--	--	--	--	--	--	--	--	--
UA	--	--	--		--	--	--	--	--	--	--	--	V	V	V
AM	--	--	--	--		--	--	--	--	--	V	--	V	V	V
AZ	--	--	--	--	--		--	--	--	--	V	--	V	V	V
GE	--	--	--	--	--	--		--	--	--	V	--	--	--	--
KZ	--	--	--	--	--	--	--		--	--	V	--	V	V	V
KG	--	--	--	--	--	--	--	--		--	V	--	V	V	V
TJ	--	--	--	--	--	--	--	--	--		V	V	V	V	V
TK	V	V	V	V	V	V	V	V	V	V		V	V	V	V
UZ	--	--	--	--	--	--	--	--	--	V	V		V	V	V
LV	V	V	V	V	V	V	V	V	V	V	V	V		--	--
LT	V	V	V	V	V	V	V	V	V	V	V	V	--		--
EE	V	V	V	V	V	V	V	V	V	V	V	V	--	--	

Notes: All information for regular passports; -- = No visa required; V = visa required.
© Irina Ivakhnyuk/T. Savvidis

According to the *United Nations*, there were globally about 214 million people living outside their place of birth in mid-2010, which is about three per cent of the world's population.[2] More than a third of these belong, according to *World Bank* estimates for 2006-2008, to the combined migration systems of the CIS space and Eastern Europe (Abazov 2009: 1f.).

There are two seemingly contradicting tendencies in international migration of today: On the one hand, economic globalization has brought about the internationalization of markets, including labor markets. On the other hand there is an increasing trend of **regionalization** of international migration. Since the 1980s the regional share in international migration increased from 20% to 50%. At present, half of all cross-boundary migrants stay in their native region, migrating rather to adjacent than remote countries. As a result of the regionalization in international migration, regional migration systems emerged, although often overlapping. As examples for overlapping migration systems we may quote the European or EU regional migration system and the above mentioned CIS or Eurasian system[3], in particular with regard to the migration of highly skilled persons. The CIS or Eurasian migration system is the second largest by its quantity system after the North American, or US regional migration system. The total number of foreign-born residents in the CIS zone is estimated between 25 and 30 million people, including 13 million in the Russian Federation (or 8.7% of the population in 2010)[4], seven million in the Ukraine, three million in Kazakhstan, one million in Uzbekistan, 0.5 million in Belarus, etc.[5]

The specifics of migration inside the Commonwealth of Independent States can be best described as a system of complementarity, based on the following facilitating factors:

- Historical ties for at least one and a half centuries (Central Asia) and two centuries in the case of the South Caucasus;
- Geographical proximity, 'transparent' borders (no visa-regime);
- Common transportation infrastructure and relatively low travel, allowing frequent and unlimited movement between countries of origin and host countries;
- Psychological easiness to move (Russian language as *lingua franca*, former common territory; familiarity with the mores and customs of the host country, including common patterns of resolving conflicts): "Most of the people in the CIS zone who entered schools before 1991 speak the Rus-

2 http://esa.un.org/migration/p2k0data.asp

3 The term was coined by Irina Ivakhnyk.

4 Data of IOM – cf. http://www.iom.int/jahia/Jahia/russian-federation

5 "Estimates vary widely and place the migration figures anywhere between 2.1 million (2006, CIS Statistical Committee) and 10 million people (2009, Russian Federal Migration Service) to about 15 million people (2006, World Bank Estimates)." (Abazov 2009: 2, footnote 5)

sian language and display remarkable similarities in cultural preferences, work ethics and attitudes towards team-work and conflict management" (Abazov 2009: 20-21);
- Demographic complementarity;
- Mutual interest towards common labor market;
- Large-scale irregular migration;
- Regional cooperation aimed at coordinated migration management (Ivakhnyuk 2006: 1-2).

In other words, complementarity in this context describes a situation when the involved sending and receiving countries gain from the migration exchange. Low-wage countries with high rates of unemployment and underemployment discharge their labor markets and domestic policies by exporting surplus labor, while the receiving countries fill the deficits of their labor forces and compensate their aging native populations. According to the UNDP, in 2007 and even before the economic crisis of 2008-2010, Armenia had an estimated unemployment rate of 33%, and Georgia of 35-40%. In comparison with both countries, the Russian labor market holds a bar far lower degree of competition: While on average one hundred persons apply for one post in Armenia and 30 in Georgia, it is less than three applicants in Russia.

The disparity of incomes in the countries of origin and the host countries are another major pull factor, causing massive emigration in particular during the 1990s. "A steep decline in real personal incomes and wages (…) led to the rise of extreme poverty in some republics of the CIS zone, especially in the so-called southern belt – Kyrgyzstan, Tajikistan, Armenia, Moldova and Uzbekistan" (Abazov 2009: 16). According to CIS statistics for 2008, the Russian Federation was leading with an average monthly wage rate of 718 USD, followed by Kazakhstan ($485), Belarus ($396), Ukraine ($356), Azerbaijan ($317), and Armenia ($293), while Moldova ($245), Kyrgyzstan ($137) and Tajikistan ($63) tailed the rating (Karimov 2008). In comparison, the average wage in Russia was nearly 2.5 times higher than Armenia. However, wage and income difference are very uneven, when comparing the situation in urban and rural areas: According to the CIS Committee's report a big difference in living standards between capital cities and province is traced in every country of the Commonwealth. Therefore, the average wage in Moscow is 1.9 times higher than in other cities of the country.

The near-disappearance of social welfare and the severe decline of public healthcare systems in the 'southern belt' of the CIS space are additional major push factors. "Anecdotal evidence suggests that in some remote areas, like mountainous regions in Kyrgyzstan, Tajikistan, Uzbekistan, Georgia and Russia, these welfare and healthcare systems have collapsed altogether" (Abazov 2009: 16).

There are financial gains from migration for both the sending and the host countries: For the sending CIS states these are the direct money transfers from abroad, which for example in Armenia amount to 10-20% of the GDP. An increase of income levels in the countries of origin as a result from international migration had also been observed. Host countries gain additional tax profits and increase their economic competiveness by wage cuts due to the competition of migrants in the labor market. They also gain political stability in neighboring countries, if the national labor markets there are relieved from the surplus workforce.

However, the complementarity of the Eurasian migration system has its limitations. The academic and more so the public discourse emphasize the financial loss that the money transfers to countries of origins represent. As a major receiving country, Russia for example is the main source of remittances sent to other CIS states. According to the Central Bank of Russia, the total amount of these money transfers increased by seven times between 1999 and 2004, i.e. from 0.5 billion to 3.5 billion USD. According to the national Bank of Kazakhstan, since 2000 the remittances sent by official channels from that country were growing 1.5-2 times annually, and by 2005 exceeded one billion USD (Sadovskaya 2006). Whether the financial losses that a CIS state suffers by international money transfer of migrants is compensated by income taxes and other state revenues depends largely on the ratio of irregular migrants, which is generally believed to increase under conditions of economic crisis. For Russia, the expert estimates of the number of irregular migrants range from 5-6 million.[6]

For the sending countries, losses caused by massive emigration are usually measured as losses of human capital ('brain drain' and 'brain waste'). They are believed to cause also demographic gender and generational imbalance and thus accelerate the general trend of 'aging societies'. Of all CIS states, Armenia has the highest share of its workers abroad – perhaps 700,000 (= 58%) of a labor force of 1.2 million, followed by Moldova (700,000 migrants in a labor force of 1.5 million) and Azerbaijan (up to 1.5 million migrants in labor force of 3.8 million).[7] Comparing the top 29 emigration countries of 2010, Armenia and Georgia were on place 17 and 20, respectively, with 28.2% (= 870,200 emigrants) and 25.1% (= 1,057,000) of their total populations being abroad. In 2000, the emigration rate of the tertiary-educated population was 8.8% for Armenia and 1.6% for Georgia (World Bank 2011: 61, 122). Since effects of 'brain drain' largely depend on the population size and the education level, the negative effects are expected to be greater on small countries (lesser than 30 mln), especially "such countries as Armenia, Georgia, Kyrgyzstan, Moldova and Tajikistan" (Mansoor & Quillin 2007: 183). During the peak years of post-Soviet out-

6 Data of IOM – cf. http://www.iom.int/jahia/Jahia/activities/europe/eastern-europe/russian-federation

7 Russia: CIS Migrants. „Migration News", July 2006, Vol. 13, No. 3 - http://migration.ucdavis.edu/mn/more.php?id=3209_0_4_0

migration, 1991-95, Armenia saw a decrease in scientists involved in research from 15,000 to 3,000 (Mansoor & Quillin 2007: 184).

The post-Soviet Eurasian migration system can be tentatively divided into migratory and possibly competing **sub-systems**: Three large and resource-rich territorial post-Soviet states attract labor migrants from adjacent resource-poor and low wage countries with underdeveloped and insufficient labor markets, slow or even reverse employment growth and high competition on the national labor markets. These three major receiving countries in the post-Soviet space are first of all the Russian Federation, followed by Ukraine and Kazakhstan, while the likewise resource rich CIS members Azerbaijan and Turkmenistan do not attract migrants. Before the international financial and economic crisis hit the post-Soviet space, Kazakhstan had developed into an immigration alternative for migrants from the other Central Asian republics, while Ukraine became an alternative for migrants from Georgia, who since the so called Rose Revolution of 2004 suffered increasingly from the deteriorating foreign relations between Georgia and Russia. As a host-land for Central Asian migrants Kazakhstan, on the other hand, compensated labor deficits suffered due to the loss of a fifth of its population that massive emigration of Slavic people, Germans and Jews had caused earlier in the 1990s (Ivakhnyuk 2006: 3).

With regards to demographic criteria such as population growth and generational stratification, the CIS space can be subdivided into three groups: a) The predominantly Muslim Azerbaijan and four of the five Central Asian republics (Uzbekistan, Turkmenistan, Tajikistan, Kyrgyzstan) are characterized by a moderate population growth, a high ratio of young people (29-34% of the population younger 14 years) and a significant part of the population that entered the labor market in the 1990s and 2000s. b) The second group comprises Belarus, Moldova, the Russian Federation and the Ukraine and has been characterized for about two decades by quite low fertility rates at about 1.3 to 1.6 children per woman. As of 2009, in these countries only about 14 percent of the population have been younger than 14 years of age and the proportion of people older than 60 is increasing rapidly. c) The third and medium group includes Armenia, Georgia and Kazakhstan. In these countries, the fertility rate remains high enough to maintain a stable population or to allow even a small natural population growth (Abazov 2009: 5-6).

3 Socio-Economic Push Factors of Massive Migration from Armenia and Georgia

The mobility of residents in the post-Soviet space has been less triggered by the new freedom of movement, but was rather forced upon the people by the profound socio-economic changes that followed the free-market oriented reforms, induced by the new and economically inexperienced elites after 1991. "The reforms included deregulation of the public sector and labor markets, price liberalization, mass privatization and support of private initiatives. These policies implied that the states should trust the self-regulating forces of the market to regulate the survival and failure of enterprises, and that the governments and state institutions should not directly intervene into the economic development. However, this approach combined with the policy mistakes led to a decade-long economic recession, which brought whole sectors of economy to the total collapse in many CIS countries with the partial exception of Belarus, Turkmenistan and Uzbekistan" (Abazov 2009: 11).

As a result, between 1989 and 2002 Georgia's real GDP declined 78 percent, which was the maximum decline in all post-Soviet states, while Armenia's real GDP declined 63 percent and Azerbaijan's 60 percent. Grave policy mistakes in response to consumer goods shortages, large output losses and increase in unemployment have contributed to the deterioration of the economic situation in the CIS zone. Scholars cite the "[tight] export controls" that caused "short-term disruption of trade patterns" between the CIS countries, as the new independent countries introduced new currencies, new legal arrangements for import and export operations, strict customs control on goods and services and passport control on the movement of people. With practically nonexistent coordination in the legal and economic reforms, these arrangements often led to the creation of additional and serious barriers for trade and economic cooperation and contributed to further economic decline that ultimately climaxed in the prolonged economic recession and the rise of mass unemployment and underemployment (Pomfret 1995: 43, 51-53). In addition, there were the negative economic repercussions of globalization, such as tough competition from cheap products from developing countries. Imports from China, India, South Korea and Turkey hit in particular hard the republics with relatively large labor-intensive textile and garment industries (Kazakhstan, Kyrgyzstan, Tajikistan, Azerbaijan, Georgia and Moldova) (Abazov 2009: 12).

The development during the 2000-2008 period was characterized by some macroeconomic stabilization and economic growth, though the pattern of investments and growth-rate has varied from country to country and even within the countries, with disproportional concentration of investments and economy in the largest metropolitan centers. At the same time the already existing discrepancies between the resource-rich and resource-poor post-Soviet countries increased, generating new jobs in the first category, while Armenia, Kyrgyzstan, Tajikistan, Moldova and Uzbekistan struggled to revive their economies and to generate new jobs. Although belonging to the resource-poor countries, Georgia that earlier had faced a steep economic decline in the 1990s attracted substantial foreign direct investments and aid due to its 'Rose Revolution' (2004) between 2005 and 2008.

As a result of the profound recession of the 1990s, extreme poverty (living on less than one USD per day) which at that time declined in all other parts of the world unexpectedly grew to high dimensions in the Eastern European and Central Asian parts of the post-Soviet space. Although poverty rates declined from 2000-2008, a 2002-2005 OSCE household survey in Armenia revealed 33.3% of the economically active population as unemployed (Minasyan et al. 2006: 13)—against 24.8% in 2005-2007 (Minasyan et al. 2007), while the official poverty rate was 32% in 2006.

The highest number of unemployed is found in the age group of 35-55 years. Youth unemployment is as high as 21%, and the number of female unemployment rate is more than three times higher than that of men. Long-term unemployed represent significant parts of the unemployed in Armenia, as low or unskilled workers. Lack of transportation, numerous children and difficulties in supervision of children and domestic violence also restrict the employment opportunities of unemployed (USAID Armenia 2008: 9).

As late as 2005 and under foreign pressure did Armenia start to more systematically explore the labor market situation. Since then, four empirical studies have been conducted, the results of which are research by the USAID *Social Protection Strengthening System* (SPSS) (USAID Armenia 2008: 9). Among other things, Armenia still lacks a national employment plan. A 2007 study by the World Bank drew the conclusion that employment agencies and programs in Armenia are rather limited and underfunded (USAID Armenia 2008: 6). Furthermore, Armenia has not yet been able to reintegrate the unemployed into the labor market. Similarly, there is a lack of experience or programs for self-employment or to qualify unemployed for the establishment of small companies, or offers of internships and training opportunities for young people. Neither educational institutions, nor employers are prepared or have the skills to professionally qualify employees, trainees or students to be competitive on the national and international labor market (USAID Armenia 2008: 6).

In 2008 the minimum monthly wage in Armenia was 25,000 AMD (≈ €55), the average salary 88,942 AMD (≈ €200). The highest incomes are achieved in the financial industry (on average €600 per month), followed by the construction and property industry, the transportation and communication sector and the service sector. The tax rate for salaries over 80,000 AMD (€180) per month is 20% and 10% for lower incomes (bw-i 2009: 13). Since a significant proportion of the working age population works in the informal sector, the treasury faces a correspondingly high failure of income tax.

As of the end of July 2010, the State Statistical Office of Armenia (ArmStat) calculated the monthly minimum basket (Consumer costs) with 41,052 AMD (≈ €82.38) and the minimum cost for food with 26,485 AMD (≈ €53.15). Compared with the 2nd Quarter of 2010 consumer costs increased in the 3rd Quarter of 2010 by 6% to 43,500 AMD (≈ €87.29) or by 5.6% to 28,000 AMD (≈ €56.19) for food. Social transfers in Armenia are insignificant, leaving a large gap to the actual needs of the poor.

The question of how many people in Armenia must be classified as poor or needy is answered differently in the literature, with variations between a third to a quarter of the total population (the latter especially for the period 2006-2008) below the poverty line. The above mentioned USAID-funded study concludes, inter alia, that the assumption of a poverty reduction in 2000-2003 is incorrect. Rather, the number of families on social assistance had been reduced by one-quarter (Martirosyan 2008: 3). A more general conclusion in the USAID Armenia Study sounds: "Despite expenditures reaching 64 million AMD in 2007, the existing government programs for social transfers have not contributed to the mitigation of social tension and the reduction of poverty in the Republic of Armenia. Additionally, a lack of clearly defined socio-economic and political objectives, along with inadequate development of monitoring and supervision systems, not only hinders combating extreme poverty, but also does not create guarantees for overcoming it during coming years" (Martirosyan 2008: 21).

The United Nations come to even more critical conclusions: "(...) about one quarter of population in Armenia was still poor in 2007 and about 120,000 people suffered from inadequate daily calorie intake. The 2007 Household Income Expenditure Survey (HIES) has revealed that the depth and severity of material poverty increased, suggesting that in the future it will be more difficult to reach those who are below the poverty line. The current poor are those who benefited very little, if at all, from economic growth and enhanced social assistance of the last eight years. (...)

To date, poverty reduction outside the capital has relied significantly on fiscal transfers and private remittances and this will likely decrease during the period covered by the UNDAF due to the global economic slowdown that will impact those Armenians who work overseas (in particular in Russia). Data also shows that poverty rates are consistently higher in *marzes* [provinces; TS] that are at risk of earthquakes or are in regions with unfavorable agricultural conditions and a lack of basic infrastructure, as well as those who live in border regions of the country. All told, around 36 percent of the population live in rural areas and are engaged in subsistence agriculture and have income levels far below the per capita average for the country. Similarly, people living in small and medium towns—characterized by limited employment opportunities and a failure to attract new investment—have seen only a marginal decline in poverty rates. This suggests that a strong indicator of vulnerability is the region of residence and that there is a need to urgently focus on reducing regional disparities" (UNDAF 2010: 1).

However, there is no automatism between poverty and migration rates in one region. According to data of the Armenian Statistical Office for the 2003-2006 periods the average poverty rate in the eleven administrative units of Armenia - *marzer* – was approximately 30%, while emigration rates vary considerably. With only 12.2% the emigration rate is unexpectedly low for Armenia's poverty stricken Shirak region where the aftermath of the 1988 earthquake is still felt. Shirak holds the highest amount of poverty in the country (42.9%). The capital Yerevan, on the other hand, has an average for Armenia poverty rate of

30.3% against an emigration rate of 24%, which is the highest local migration rate in the country. In the case of Yerevan, general metropolitan mobility and the presence of an urban middle provide additional explanations for large scale emigration:

Figure 1: Average share of poverty and emigration by region, 2003 to 2006*

*Poverty percent average is for the 2004-2006 period.
Source: *Marzes* of the Republic of Armenia in Figures, 2001-2005; *Marzes* of the Republic of Armenia in Figures, 2002-2006; and Social Snapshot and Poverty in Armenia, 2007. www.armstat.am

Finally, there exist additional regional specifics that cause the under-financing of social, health and education resorts and the disproportionately high financing of the defense budget in all three South Caucasian republics: Ethnic and territorial conflicts inflate the expenditure on weapons beyond all dimensions and on the expenses of all other demands in state and society. In this 'race of arms and budgets', Azerbaijan has the lead, followed by Georgia.

Table 2: Annual expenditure on defense in the state household of South Caucasian states 2004-2008

Year/Country	Armenia	Georgia	Azerbaijan
2008	US $376mln	US $769mln	US $1.3 billion
2007	US $280mln	US $310mln	US $1 billion
2006	US $166mln	US $218mln	US $700mln
2005	US $136mln	US $180mln	US $300mln
2004	US $81mln	US $60mln	US $175mln

4 De facto Refugees, Internally Displaced Persons and Asylum Applicants from Armenia and Georgia

According to the United Nations, there were more than 16 million refugees worldwide in the mid-year of 2010.[8] Among international migrants from Armenia we find an extraordinary high percentage of refugees, even two decades after the events that caused the massive flight of ethnic Armenians from the territory of Azerbaijan, Nagorno Karabakh and the Armenian borderlands adjacent to Azerbaijan. This high ratio is indicative for the failure of the Armenian state to economically and socially integrate the inflow of refugees from neighboring Azerbaijan, who arrived in Armenia between 1988-1994 during a period of extreme economic and political crisis, caused by the combined effects of a disastrous earth-quake (1988), the blockade of land roads and railways by Azerbaijan (since 1989) and Turkey (since 1993), and the near to complete collapse of Armenia's industries in the early 1990s, when the land blocked country was nearly cut off from Russian supplies of natural gas and oil. In addition, the newcomers from Azerbaijan frequently did not fit into the profile of the Armenian labor market, in particular not in the period in question. Many were of urban and industrial background – from the capital city of Baku, or the industrial city of Sumgait. The refugees had been professionals and skilled workers in Azerbaijan's petro industries, but were unable to adapt to the conditions of a subsistence economy in which people barely meet their everyday needs. With the predominantly rural Azeri minority of Armenia expelled or gone, the urban refugees from Azerbaijan were given the houses and farms of Armenia's Azeri minority, but most refugees would not know how to milk a cow, to grow potatoes or grain. In this misery, emigration, temporary or seasonal migration often appeared as the only alternative.

The *United Nations Convention Relating to the Status of Refugees* (1951) defines a refugee as a person who according to the formal definition in article 1A of this Convention, is "owing to a well-founded fear of being persecuted for reasons of race, religion, nationality, membership of a particular social group, or political opinion, is outside the country of his nationality, and is unable to or, owing to such fear, is unwilling to avail himself of the protection of that country". This concept was expanded by the *Convention's* 1967 *Protocol* and by regional conventions in Africa and South America to include persons who had fled war or other violence in their home country. In the UN classification, refugee women and children represent an additional subsection of refugees that need special attention.

The first post-Soviet decade (1991-1999/2000) has been dabbed as the phase of ethnic migration, peaking between the years 1993-1996 when the migration was mainly driven "by ethnic, cultural and education changes" and the migrant groups consisted predominantly of people who decided to return perma-

8 http://esa.un.org/migration/p2k0data.asp

nently to their titular homeland states (Abazov 2009: 17). The inability of the inexperienced post-Soviet administrations to accommodate and integrate incoming refugee communities was an additional 'driver' and frequently caused remigration. Estimates range from nine to fifteen million 'ethnic migrants' (Abazov 2009: 21)[9] with roughly three million from the South Caucasus alone. The dissolution of the USSR and the emergence of new post-socialist states led to an enormous increase of refugees and expelled people who were fleeing from ethnic strife and even war, deterioration of minority rights or the anticipated violation of such rights by the new, often nationalist motivated legislation of the post-Soviet states. Important, but less obvious reasons for the massive migration were social marginalization and the "removal of the elaborate system of ethnic, gender and social quotas" that existed during Soviet times. "Many migrants have taken into consideration not only ethnic and cultural factors, but also social and economic factors in calculating cost benefits of such migration" (Abazov 2009: 22).

In the South Caucasus the hot spots of ethnic strife and wars were identical with the neglected and disadvantaged regions of Abkhazia, South Ossetia and Nagorno Karabakh that had been annexed to Georgia in the early 1920s against the will of their nominal populations (Abkhazia and South Ossetia) or, in the case of Nagorno Karabakh, seceded to Azerbaijan despite the resistance and continuous protest of Karabakh's absolute Armenian majority. The armed conflicts on and in these areas, together with frequent reciprocal atrocities among the conflicting ethnic groups caused the massive flight of local majorities and minorities. In the case of Abkhazia, it was predominantly the ethnic Georgian population that fled the area during and after 13 months of war and killings: Four fifths of a total of approximately 250,000 refugees from Abkhazia were ethnic Georgians. Most of these - 46% of all refugees from Abkhazia (Selm 2005) - escaped to the neighboring Georgian province of Samegrelo-Zemo Svaneti, while the others were dispelled over various regions of Georgia. There is a significant trend of internal migration in Georgia, especially to the metropolitan capital Tbilisi where 29.6% or the second largest group of Georgia's IDPs is residing. The failure of Georgia's administration and decision-makers to efficiently provide these IDP communities with employment and housing even 17 years after the Abkhazian-Georgian war causes an above-average susceptibility to migration among them.

After her abortive 2008 war over South Ossetia, Georgia received substantial foreign financial aid to accommodate and integrate refugees not only from that area, but in general. The one-sided attention to ethnic Georgian refugees from South Ossetia and the neglect of the refugees from Abkhazia, who felt se-

9 Relating to Andrey Korobkov, R. Abazov mentions 710,000 'forced migrants' in the Post-Soviet space as of 1991. 'Forced migration' peaked in 1998 with 1.2 mln persons registered by the statistical agencies (Abazov 2009: 29-30).

cond class, caused much public protest in Georgia's refugee community. After nearly 17 years of indolence and indifference, Georgia has now pledged to actively address to the problems of refugees from Abkhazia and to provide accommodation until the year 2013. However, until today such attempts often remain insufficient, since the newly built refugee settlements are located in remote and structurally weak areas where the new settlers remain unemployed (Gvalia 2011). IDPs, who were evicted in January 2011 from the capital Tbilisi also complained about the systematic lack of basic utilities and occasional sub-par building conditions (Corso 2011). Furthermore, Georgia's settlement policy creates tensions between indigenous residents and incoming refugees. It is for all these reasons that refugees, or IDPs from Abkhazia cling to the capital city Tbilisi where they hope for better employment opportunities.

Since February 1988, the increasingly violent conflict on and in the Autonomous Region of Nagorno Karabakh had uprooted approximately 340,000 ethnic Armenians in Azerbaijan and about 200,000 Azeris in (Soviet) Armenia. Azerbaijan's abortive attempt to militarily re-conquer the break-away region during December 1991-May 1994 caused the flight of another 80,000 ethnic Armenians from the afflicted area. Whereas most of the latter returned after an Armenian victory in 1994 despite the fact that so far no peace or conflict agreement could be achieved, many of the Armenian refugees from Azerbaijan could not be embedded in post-Soviet Armenia, from where they left for Russia and third countries.

In Armenia, 40,000 (about 5%) of the approximately 800,000 families are left without permanent accommodation. About 40% of these (2% of the total number of households) are living in provisional container-like accommodation ("Domik" - "little house") in the northern Armenian earthquake zone, up to ten percent more in residential homes and defective, unsafe and therefore in danger of collapsing houses. Others live in former hotels, schools and kindergartens that were converted to emergency accommodations. In addition to the homeless another estimated 70,000 families are in dire need of better housing. About 20,000 of the homeless families are considered as socially extremely vulnerable. Since investment by the government and foreign donors aims primarily at this group the housing needs of the likewise vulnerable middle class families remain unattended (Social Housing 2010). As these figures suggest, homelessness and lack of proper accommodation are not only basic problems of refugees and IDPs, but they hit these socially vulnerable cohorts disproportionately. As Armenian media reported, only in April 2010 did 76 refugees from Azerbaijan receive new apartments, built at the expenses of the Japanese government. Since 1988, these refugees had lived in the village of Kazakh (Province of Kotayk) in a former prison building.[10]

10 Cf. http://www.armenianow.com/hy/node/22084

Among the refugees, who left the South Caucasus in the first half of the 1990s is a significant share of Russophones whose rights and privileges had been threatened by the new language laws that were adapted in Georgia (1992) and Armenia (1993), making the language of the nominal population the only state language of the new independent countries. In Armenia and also in Azerbaijan, the linguistic minority of Russophones comprised parts of the ethnic majority population, often from the capitals Yerevan and Baku and with an elitist, 'cosmopolitan' background. In Georgia, the Russophone community comprised various ethnic minorities with urban background, such as Russians, Armenians, Greeks, and Jews.

Estimates of Georgian nationals who left the country in the 1990s, range between 300,000 and more than 1.5 million (Selm 2005). While the ethnic phase of late and Post Soviet emigration from the South Caucasus seems terminated, - between 2000 and 2009 the proportion of ethnic migrants from the post-Soviet space declined to about five to ten percent (Abazov 2009: 23), - emigrants from Armenia and in particular from Georgia continuously apply for political asylum in third countries. The number of asylum applications by Georgians has increased steadily since 2000, when there were only 3,998 applications in industrialized countries.

According to the statistics of the United Nations the percentage of refugees among the international migrants from Armenia remains high, with the peak of 50.2% in 2000. But even ten years later nearly every fifth migrant from Armenia – 18.2% - was officially recognized as a refugee, while the results for Georgia show that there were nearly no refugees among the international migrants from that country (cf. Table 3):

Table 3: Armenia (1990-2010) – Refugees as a percentage of international migrants

Year	Refugees as a percentage of international migrants
1990	--
1995	38.4
2000	50.2
2005	46.2
2010	18.2

© United Nations: http://esa.un.org/migration/p2k0data.asp

Table 4: Georgia (1990-2010) – Refugees as a percentage of international migrants

Year	Refugees as a percentage of international migrants
1990	--
1995	--
2000	2.9
2005	1.3
2010	0.7

© United Nations: http://esa.un.org/migration/p2k0data.asp

However, a different picture is revealed if numbers of asylum applicants are compared: According to UNHCR, in 2003 there were some 9,500 asylum applications made by Georgians in other countries, of which more than 7,000 were made in the then EU-15 Member States, particularly in Austria, France, and Germany. A further 1,300 applied in states that joined the EU in May 2004. This development so far peaked in 2009, when 11,000 nationals of Georgia applied for political asylum abroad and Georgia moved from the 21st place of source countries of asylum-seekers (2006) in 44 selected industrialized countries to the 10th place (2009) in only four years (UNHCR 2010: 12). While Georgia saw a significant increase (+102%) of asylum applications, the number of applications by asylum seekers from Armenia and Azerbaijan during the same time increased by 43%, and 36% respectively.

Table 5: Asylum applications - Changes in the ranking of the top-10 countries of origin*

Country/place of ranking	2005	2006	2007	2008	2009
Afghanistan	9	6	8	4	1
Iraq	4	1	1	1	2
Somalia	11	8	6	2	3
Russian Federation	2	3	2	3	4
China	3	2	3	5	5
Serbia	1	4	4	6	6
Nigeria	8	13	13	7	7
Iran	6	5	9	11	8
Pakistan	10	9	5	8	9
Georgia	14	21	20	17	10

* Statistics prior to 2007 refer to Serbia and Montenegro.
© UNHCR (2010)

While some South Caucasian applicants in industrialized host countries are given refugee status on the grounds of their individual case, the vast majority apparently applied, because this was the only way to gain a regular status of residence in a country of destination. The increase of applications by 2009 seems to reflect, on the one hand the general deterioration of the socio-economic situation during the global financial and economic crisis. In comparison with neighboring

Armenia and Azerbaijan the twice or even thrice times higher number of Georgian applications, on the other hand, may be explained by Georgia's special situation with regard to her relations toward Russia. The cancellation of visa for the average tourist from Georgia prevents entry to Russia which until the brief war of August 2008 had been the main receiver of Georgia's surplus workforce. Such limitations compel Georgian labor migrants to look out for employment and residence opportunities in other industrialized countries, preferably in the EU space.

Perhaps the market radicalism and **economic libertinism** adopted by the government of President Mikheil Saakashvili could be quoted as additional reason that drives Georgian nationals out of the country. In 2006, Georgia issued a new law on labor that has been qualified as one of the most anti-labor laws in the world, and in autumn 2010 Saakashvili vowed to make Georgia "a flagship of the global economic liberalism", among others with the new *Law on Economic Freedom, Opportunity and Dignity* and several constitutional amendments (Jobelius 2010: 3). With employment and trade union rights undermined by this law, Georgia is in conflict with ILO core labor standards and the *European Social Charter* (1961, revised 1996) (Jobelius 2010: 1).

5 Migration Profiles in Comparison

5.1 General Trends and Gender ratio

The United Nations give the following migration country profiles for Armenia and Georgia:

Table 6: Armenia 1990-2010 - Migration profile

Indicator	1990	1995	2000	2005	2010
Estimated number of international migrants at mid-year	658,789	681,557	574,235	492,570	324,184
Estimated number of refugees at mid-year	0	261,495	288,404	227,393	59,140
Population at mid-year (thousands)	3,545	3,223	3,076	3,065	3,090
Estimated number of female migrants at mid-year	388,254	401,673	338,423	290,294	191,056
Estimated number of male migrants at mid-year	270,535	279,884	235,812	202,276	133,128
International migrants as a percentage of the population	18,6	21,1	18,7	16,1	10,5
Female migrants as percentage of all international migrants	58,9	58,9	58,9	58,9	58,9
Indicator	**1990-1995**	**1995-2000**	**2000-2005**	**2005-2010**	
Annual rate of change of the migrant stock (%)	0.7	-3.4	-3.1	-8.4	

Source: United Nations, Department of Economic and Social Affairs, Population Division (2009). *Trends in International Migrant Stock: The 2008 Revision* (United Nations database, POP/DB/MIG/Stock/Rev.2008)

Table 7: Georgia 1990-2010 - Migration profile

Indicator	1990	1995	2000	2005	2010
Estimated number of international migrants at mid-year	338,300	249,900	218,600	191,220	167,269
Estimated number of refugees at mid-year	0	50	6,400	2,528	1,210
Population at mid-year (thousands)	5,460	5,069	4,745	4,465	4,219
Estimated number of female migrants at mid-year	190,206	141,112	124,389	109,084	95,496
Estimated number of male migrants at mid-year	148,094	108,788	94,211	82,136	71,773
International migrants as a percentage of the population	6,2	4,9	4,6	4,3	4
Female migrants as percentage of all international migrants	56,2	56,5	56,9	57	57,1
Refugees as a percentage of international migrants	0	0	2.9	1.3	0.7

Indicator	1990-1995	1995-2000	2000-2005	2005-2010
Annual rate of change of the migrant stock (%)	-6.1	-2.7	-2.7	-2.7

Source: United Nations, Department of Economic and Social Affairs, Population Division (2009). Trends in International Migrant Stock: The 2008 Revision (United Nations database, POP/DB/MIG/Stock/Rev.2008)

According to the figures of Tables 6 and 7 the **share of international migrants in the country's overall population** is 2.5 times higher in Armenia than in Georgia: In 2010 it comprised 10.5% in Armenia against 4% in Georgia. While there was a gradual and slow decline of the migrant ratio in the overall population in Georgia – starting with 6.2% in 1990 – the trends in Armenia were much more dramatic, starting with 18.4% (1990) and peaking in 1995 with 21.1% (against 4.6% in Georgia in the same year). Nevertheless, over a period of 15 years, there was a remarkable decline of the migrant share of more than 50% from 1995 (21.1%) until 2010 (10.5%). This sharp decrease of migration shows in particular in the 2005-2010 margins with an **annual rate of change in the migrant stock** of -8% in the case of Armenia, while since 1995 the rate is a persistent -2.7% for Georgia. The near to three times higher change rate in Armenia obviously reflects the involuntary decline of migration, as caused by the 2008-2010 economic crisis, whereas the decade of 1995-2005 brought a certain stabilization for Armenia's out-migration, expressed in migration stock changes of -3.4 to -3.7%. Yet with every tenth resident of the country in migration, the share of migrants among Armenia's overall population remained still quite high in 2010.

A different picture emerges if **frequencies and duration of migration stays** are compared. According to the ArGeMi survey, the out-migration frequency is higher in the samples of migrants from Georgia than in those from Armenia: Since 1990, more than one fifth of the out-migrants from Georgia to other destinations than Moscow had stayed abroad more than four times and with a duration of more than a month (out-migrants from Armenia - only 13.3%). As to the migration to Moscow, the frequency of stays there was more than twice higher among migrants from Georgia - 19.5% of this sample had been more than four times to Moscow, while this applies to only 9% of the respondents from Armenia. According to the data of a 2006/7 study by the *European Bank for Reconstruction and Development* every second respondent (51%) from Georgia had lived in Russia more than four years, which was the highest share in long-term residence compared with migrants from Azerbaijan and Moldova.

According to various estimates the total number of émigrés from Armenia between 1988 and 2003 ranges from 800 000 to 1.5 mln, or a quarter of the total population of 3 million in 2001. Out of these people, 82.2% left in search of jobs and the rest – for family re-unification (Abdurazakova 2010: 4-5). Ninety per cent of **labor migrants** from Armenia left for other post-Soviet countries (mostly to Russia), and the rest – to the EU and USA; about half of the labor migrants from Armenia emigrated permanently (Gevorkyan et al. 2008).

In international migration, an increasing trend of **feminization** has been observed. As in mid-2010, nearly 200,000 women from Armenia and nearly 100,000 from Georgia were among the trans-boundary migrants from these countries. But contrary to Georgian expert opinion that claims the share of female migrants among the overall migration stock as particularly high for Georgia, the comparison with Armenia reveals an even slightly higher percentage: 58.9% (Armenia) against 57.1% in Georgia. In contrast to the international trend of increasing feminization, in both countries the percentage of female migrants has remained stable during the period under scrutiny (1990-2010). In other words: For two decades, more than every second migrant from both South Caucasian countries was a woman.

Partly caused by the feminization of poverty, female emigration challenges traditional gender perceptions of the "divisions of labor and responsibilities; women become breadwinners and transnational mothers, whereas men loose their role as family providers" (Lundkvist-Houndoumadi 2010: 53). In order to (re-)gain respectability, female migrants try hard to justify their separation from family and in particular their children by sending more and/or more frequently money from abroad than men or by generally becoming the 'better migrants'.

M. Lundkvist-Houndoumadi, who interviewed female migrants in Georgia and in Athens in autumn 2008, concludes: "The narratives formed around Georgian women's migration reveal the conflicts that exist between the traditional gendered expectations from these women and their actual practices. Narratives of condemnation and admiration of the emigrated women exist alongside each other. The women's own narratives often constitute ways of avoiding the condemnation by trying to adapt the gendered expectations to their altered practices. The way the emigrated women describe their emigration is often a process by which they seek to justify their choices" (Lundkvist-Houndoumadi 2010: 51).

Migration seems socially justified, if it does not serve the personal needs and interests of a woman, but fits into the collective image of the self-sacrificing female family provider or mother as it emerged already in the Soviet period: "The hardships of combining the provisioning of a family in a shortage economy with household labor, childcare and waged labor fell most heavily on the women. Studies done on gender roles in the socialist period from the 70s to the 90s showed how women generally saw themselves as courageously and unselfishly coping with these very difficult demands. Through their struggle they gained gratification and moral superiority, which also meant a certain power in the household (...). The gender relations during that period were characterized by the notions of the female 'brave victim' and the male 'big child'. (...) The female 'brave victim' was defined in contrast to the image of her husband, the socialist man, who might have been better paid, dominant at work, but who acted as the 'big child' in the family; 'disorganized, needy, dependent, demanding to be taken care of and humored, as he occasionally acted out with aggression, alcoholism, womanizing or absenteeism' (...)" (Lundkvist-Houndoumadi 2010: 62).

Admittedly, the female share in migration stocks varies considerably, first of all in dependence from the host countries, the demands of labor markets and further determinants. A near to absolute female share (96%) was found among the 12-13,000 labor migrants from Armenia to Turkey, as a 2009 survey by the *Eurasia Partnership Foundation* revealed. As a result of Turkey's economic growth, an urban middle class emerged, and the number of employed women of Turkish nationality increased; these women not only have a need on assistance in their households, but also the financial means to pay for aid. Due to Turkey's refusal to enter diplomatic relations with Armenia and the subsequent lack of diplomatic representations in both countries, employees from Armenia go to Turkey undocumented. Their status is highly precarious, and their children that had been born in Turkey remain stateless and meet difficulties to register for school education. Nearly three quarters of the migrants in question work as housemaids in Turkey, and further 18% as care givers of sick and old people, typically employed in households of members of Turkey's Armenian minority of approximately 60,000 people, who live nearly exclusively in Istanbul. With their rather good school education and professional skills, these migrant women from Armenia represent a classical case of 'brain waste'.[11]

Shares higher than the average ratio of female migrants among the international migrants from Georgia have been identified for Germany and Greece (Lundkvist-Houndoumadi 2010: 52), where migrants were badly hit by the Greek financial crisis of 2009/10.[12]

In difference to Turkey, Greece or Germany, Russia – and its capital Moscow in particular – has been believed to attract predominantly male migrants, because of the physically demanding work offers there. According to the RF *Federal Migration Service* (FMS) 84% of all labor migrants to Russia are men.

11 For details, see the report by Gevorg Poghosyan and the survey's research paper by Alin Rozinian: http://epfound.am/files/epf_migration_report_feb_2010_final_march_5_1.pdf

12 Cf. also the results of the ArGeMi event monitoring 2010 in the contribution of Irina Badurashavili.

However, the data of the ArGeMi survey show a lower share of men, namely 61.5% of the respondents from Armenia and 56.5% of those from Georgia.

Female ratios were lowest among the returnees from Moscow to Armenia (32.5%) and the respondents from Georgia in Moscow (38.5). The highest shares were found among potential migrants in Armenia (50%) and Georgia (60%), which relatively corresponds with the percentages established by the United Nations. In the cohorts of returnees from all other destinations the difference to the UN data is -11.9% for the returnees to Armenia, but only -4.1 in the equivalent cohort of returnees to Georgia. In other words: Nearly every second returnee to Georgia was a woman, regardless whether she arrived from Moscow or other destinations. The comparison of cohorts from Armenia revealed that the female share is highest among the potential migrants, followed by the returnees from other destinations:

Table 8: Gender composition in the ArGeMi cohorts (2009)

ArGeMi Cohorts	Share of women (%)	Rate (%) of difference (as compared with UN percentage)
	Migrants from Armenia	
Returnees from Moscow	32.5	-26.4
Returnees from all other destinations	47.0	-11.9
Interviewed in Moscow	44.5	-14.4
Potential Migrants	50.0	-8.9
	Migrants from Georgia	
Returnees from Moscow	48.5	-8.6
Returnees from all other destinations	53.0	-4.1
Interviewed in Moscow	38.5	-18.6
Potential Migrants	60.0	+2.9

The largest deviations from the specifications of the United Nations are found in the cohorts of returnees from Moscow to Armenia and among the respondents from Georgia in Moscow, followed by the two other cohorts linked with the Russian capital city, which in all four cases may be explained by local 'male oriented' labor market conditions.

5.2 Age Groups

In Moscow, three quarters of the ArGeMi respondents from Armenia and 66.5% of those from Georgia were younger than 35 years. The higher ratio of middle-agers among the respondents from Georgia – 8% more in comparison with the respondents from Armenia – explains partly by the fact that the influx of younger age cohorts to Russia has been increasingly aggravated over the last years due to the visa restrictions for Georgian nationals. In contrast to the findings in Moscow, the surveys in the countries of origin show that more than three quarters of all interviewed returnees from Moscow (75.5% from Armenia; 78.5% from Georgia) were older than 30 years, and more than a third of all returnees from Moscow belonged to the cohort of 45-64 years. With 40.3% the highest share of young migrants (18-29 years) was to be found among the Georgian returnees from other directions.

Table 9: Distribution of age cohorts among migrants from Armenia and Georgia in Moscow (%)

ArGeMi Cohorts	18-22 Years	23-35 Years	36-55 years	56- years
Migrants from Armenia				
Interviewed in Moscow	34.5	40	25.5	--
Returnees from Moscow	18- 29 years	30-44 years	45-64 years	65- years
	26.0	38.5	37.0	--
Returnees from other destinations	18.6	37.9	36.9	6.5
Migrants from Georgia				
Interviewed in Moscow	18-22 Years	23-35 Years	36-55 years	56- years
	31.5	35.0	33.5	--
Returnees from Moscow	18- 29 years	30-44 years	45-64 years	65- years
	18.0	39.5	39.0	3.5
Returnees from other destinations	40.3	34.7	25.4	2.7

5.3 Level of education and 'brain waste'

On average, migrants from the post-Soviet space have completed a secondary or high school. The respondents of the ArGeMi survey confirm this rule: Nearly 60% of the returnees interviewed in Georgia hold a university degree and only a quarter have less than a secondary specialized level of education. Among the interviewees in Armenia those who had been in Europe or overseas were the best skilled (43.7% with higher education). Eight of ten migrants (82%) from Armenia and Georgia in Moscow have secondary or higher education.

Table 10: Education level of ArGeMi respondents (%)

<table>
<tr><th>Cohorts of ArGeMi respondents</th><th colspan="2">Completed Secondary School or Technical College</th><th colspan="2">Completed High School or Pedagogical Training</th></tr>
<tr><td colspan="5">Migrants from Armenia</td></tr>
<tr><td>Interviewed in Moscow</td><td colspan="2">43.0</td><td colspan="2">42.0</td></tr>
<tr><td>Returnees from Moscow</td><td colspan="2">52.5</td><td colspan="2">37.0</td></tr>
<tr><td>Returnees from other destinations</td><td colspan="2">51.3</td><td colspan="2">43.7</td></tr>
<tr><td>Potential migrants</td><td colspan="2">45.0</td><td colspan="2">35.0</td></tr>
<tr><td colspan="5">Migrants from Georgia</td></tr>
<tr><td>Interviewed in Moscow</td><td colspan="2">47.5</td><td colspan="2">34.5</td></tr>
<tr><td rowspan="2">Returnees from Moscow</td><td>Completed Secondary School or below</td><td colspan="2">Technical College</td><td>Completed High School or university</td></tr>
<tr><td>23.5</td><td colspan="2">19.5</td><td>57.0</td></tr>
<tr><td>Returnees from other destinations</td><td>22.0</td><td colspan="2">13.0</td><td>65.0</td></tr>
<tr><td>Potential migrants</td><td>29.0</td><td colspan="2">15.0</td><td>53.0</td></tr>
</table>

Considerate 'brain waste' is an issue for migrants from Armenia and Georgia: More than a quarter of the respondents from both countries complained that their current employment or business in Moscow did 'not at all' correspond with their level of education. For women from Georgia the 'brain waste' situation seems to be particularly urgent, because 39% of them answered the question about the correspondence of education/training and their current occupation with 'not at all', while the samples from Armenia show a near to complete gender balance in this category (26.2% men, 26% women).

5.4 Language Command

Foreign language skills are essential for successful migration, whether labor, business or education may be its purpose. In the ArGeMi questionnaire the respondents were requested to self-assess their command of the official language of their country of origin (Armenian or Georgian), of Russian, English and other foreign languages. A scale of four degrees ('fluent', 'good', 'bad', 'not at all') had been offered for the evaluation. In general, the cohorts from and in Georgia seemed to have assessed their skills of the native/official language and of Russian with more self-confidence than those from and in Armenia. Nearly three quarters of all Georgian returnees from Moscow assume that they speak Russian fluently, whereas less than every second returnee from Moscow to Armenia taxed his command of Russian as fluent.

The two Moscow cohorts of interviewees show thc lowest degree of a fluent command of the native/official languages Armenian or Georgian (59% and 66.5%) and the highest degree of bilingualism, in particular among the respondents from Armenia: 57% (47.5% in the Georgian equivalent) assessed their command of Russian as fluent, which is nearly the same ratio as in the 'fluent' margin of Armenian. The highest percentages of a fluent command of the native/official languages are to be found among the two cohorts of potential migrants in Armenia and Georgia (88% and 97%).

In all four cohorts from Georgia only very few respondents admitted that they have a poor or no command of the official language: 6.5% among the respondents in Moscow and 2% among the returnees from Moscow reported a poor knowledge of Georgian, whereas the percentage for the 'not at all' margin was nil among the two other cohorts. In an indirect way, this also indicates the decline of ethnic emigration; at the same time it reveals the increasing linguistic Georgization of the country. Interestingly, four percent of the interviewees in Moscow from near to mono-ethnic Armenia reported that they have no command of Armenian, and further 6 percent admitted a poor command of the official language of that country, which roughly corresponds with the percentage of Russophones and ethnic Non-Armenians in the society of Armenia during the first post-Soviet decade. Again, the highest percentage of a 'fluent command' was to be found among the cohort of potential migrants (88%), but in comparison with the equivalent cohort from Georgia this percentage was 9% lower.

The cohort comparison also reveals that English as an international language of communication is still not as frequent and popular among South Caucasian migrants than Russian, which remains the traditional regional language of communication. Surprisingly, the highest relative percentage of a good command of English was found among the respondents from Armenia, interviewed in Moscow: Every third in this cohort believes that he or she has a good command of English. In comparison with all other cohorts, this particular cohort also revealed the lowest percentage of those who had no command of English at all (27.5%). This truly amazing result can be interpreted in several ways. We explain it as yet another effect of brain waste and visa regimes in other migration systems than the CIS space that forces highly educated migrants from Armenia to travel to Russia instead of more appropriate for their linguistic skills destinations.

On the other hand, the cohorts of returnees from Moscow show the highest ratio of respondents who admitted to 'not at all' speak English: 51% of the returnees to Armenia and even 74.5% of those to Georgia have no command of this language. Even the cohorts of returnees from other destinations than Russia show a high degree in the 'not at all' margin: Nearly every second respondent from Georgia and 44.7% of those from Armenia is not knowledgeable in English. Finally, the results in the two cohorts of potential migrants reveal that there is no change to be expected, although the cohorts of potential migrants are generally younger: 49% of the future migrants from Armenia and even 60% of those from Georgia have no command of English at all.

Table 11: Self-assessment of official/native and foreign language skills (percentage)

ArGeMi Cohorts	Official/native language (Armenian, Georgian)				Russian				English			
	Fluent	Good	Bad	Not at all	Fluent	Good	Bad	Not at all	Fluent	Good	Bad	Not at all
MIGRANTS FROM ARMENIA												
Interviewed in Moscow	59.0	31.0	6.0	4.0	57.0	41.5	0.5	1.0	5.0	35.0	32.5	27.5
Returnees from Moscow	83.5	15.5	1.0	-	49.0	45.5	5.0	0.5	7.5	15.5	26.5	51.0
Returnees from other destinations	79.7	20.0	-	0.3	42.0	49.0	7.7	1.3	11.0	23.7	20.7	44.7
Potential migrants	88.0	12.0	-	-	48.0	49.0	9.0	3.0	19.0	11.0	21.0	49.0
MIGRANTS FROM GEORGIA												
Interviewed in Moscow	66.5	26.6	6.5	0.5	47.5	48.5	4.0	-	8.5	23.0	23.0	45.5
Returnees from Moscow	95.5	2.0	2.0	0.5	71.0	27.0	1.5	0.5	2.0	11.5	12.0	74.5
Returnees from other destinations	97.0	2.3	0.7	-	52.0	31.7	10.3	6.0	13.3	23.0	14.0	49.7
Potential migrants	97.0	3.0	-	-	28.0	41.0	19.0	12.0	13.0	16.0	11.0	60.0

5.5 Occupation

The comparison between the occupational stratification in the countries of origin and in Moscow shows no big differences with the one exception of the trade sector: 12% of the respondents from Armenia had already worked in the trade sector prior to migration, while in Moscow their share increased to 31.9% (for the respondents from Georgia the figures are 18.5% and 38.9% respectively). However, the returnees from Moscow who were interviewed in Armenia mentioned only a 12.5% share for occupations in the trade sector during their stay in Moscow, which fully corresponds with the occupation rate in the country of origin. According to the RF *Federal Migration Service*, 20% of all migrants in Russia work in the trade sector and 40% in the construction sector. The ArGeMi polls in Moscow, however, reveal much lower figures for the construction industries: Only 14.7% of the respondents from Armenia worked in the construction industries in Moscow (21% according to the poll, conducted in Armenia among returnees from Moscow). The over average representation of the Muscovite cohorts in the trade sector can be explained by specifics of the Muscovite labor market situation and also with informal employment traditions as established during the Soviet period in the retail sector, where South Caucasians were well represented (cf. chapter 2).

Table 12: Occupational distribution of migrants from Armenia and Georgia in countries of origin and in Moscow (in brackets: returnees from Moscow interviewed in countries of origin)

Occupational Sector	Migrants from Armenia		Migrants from Georgia		Overall Migrants in RF (according to FMS)
	In Armenia	In Moscow	In Georgia	In Moscow	
Construction	5% (22.5%)	14.7% (21%)	7.5% (8.3%)	13.0%	40%
Industries	4.0% (16.5%)	6.9% (5.5%)	8.0% (12.8%)	6.9%	7.0%
Transportation	5.0% (10%)	9.5% (2%)	7.0% (11.9%)	11.5%	--
Agriculture	3.5% (2.5%)	0.9%	2.5% (0.9%)	--	7.0%
Trade	12.0% (--)	31.9% (12.5%)	18.5% (24.8%)	38.9%	20.0%
Education	6.5% (13%)	6.9% (1.5%)	7.5% (9.2%)	9.2%	--
Health	5.5% (5%)	10.3% (1.5%)	2.0% (5.5%)	5.3%	--
Others	4.5% (13.5%)	19.0%	2.5%	15.3%	

5.6 Employment Assessment

Despite all discrepancies between education and occupation the overwhelming majority of employed ArGeMi respondents expressed a full or relative satisfaction with their current employment in Moscow (58% from Armenia, 62.6% from Georgia). A third of the migrants from Armenia and a quarter of those from Georgia mentioned even full satisfaction. In gender comparison the percentage of discontent with their employment situation was higher among the female respondents.

In the general cohort comparison returnees from Moscow to Georgia appear as the most content of all samples, with 88.4% of all respondents expressing full (40.4%) or relative (48.4%) satisfaction with their most recent employment in Moscow, while the equivalent sample from Armenia shows the lowest degree (42.5%) of contentment (22.5% were fully, 20% partly satisfied). However, the returnees to Armenia from other destinations showed only a slightly higher degree of satisfaction with their employment during their last stay abroad (20% were fully, 28% partly satisfied).

The discrepancies between the higher degrees of contentment, articulated by the respondents in Moscow may be explained by a) differences in the interview situation between returnees to their homelands and respondents in Moscow, who+ were interviewed not by compatriots, but by representatives of the Russian majority population. b) Differences between returnee samples from/in Armenia and Georgia probably derive from the generally critical reception of migration related issues among migrants from Armenia (cf. also 8: Migration evaluation).

5.7 Income Situation

The ArGeMi questionnaire did include questions about the respondents' personal and family income situation. Since these questions relate to current income levels, only the responses of the two cohorts interviewed in Moscow reflect the income levels during a migration stay.

In comparison with the average Muscovite population, immigrants from Armenia and Georgia seem to do quite well, if they can find employment or run their business: About a third of the ArGeMi respondents had a personal income corresponding with or even surpassing the officially established average income level in Moscow. But related to the number of 3-4 persons, who typically live in the respondents' households, the ArGeMi survey results reveal a real per capita income of only 8,500 RUB or 11,333 RUB[13], which is more than three times lower than the official per capita income estimate of 34,814 RUB (≈ €859) for migrants from Armenia and 31,263 (≈ €770) those from Georgia as given by the *Moscow Statistical Office*. The real monthly per capita income is roughly €210 in a migrant household of four and €274 in households with three members (with a mean value of €242).

---

13 Depending on whether the household consists of three or four persons

Table 13: Average monthly income of ArGeMi respondents (%)

ArGeMi Cohorts	No personal income	< $50	$51 to 100	$101 to 200	$201 to 300	> $300
MIGRANTS FROM ARMENIA						
Returnees from Moscow	27.0	7.0	12.5	23.0	30.5	-
Returnees from other destinations	23.0	7.0	12.3	27.7	29.7	-
Potential migrants	32.0	3.0	12.0	16.0	22.0	15.0

Interviewed in Moscow	No personal income	< 4,500 RUB (≈ $152)	4,500-8,000 RUB (≈ $152-269)	8,000-16,000 RUB (≈ $269-539)	16,000-24,000 RUB (≈ $539-808)	24,000-40,000 RUB (≈ $808-1,346)	> 40,000 RUB (≈ $1,346)
	24.0	5.0	3.0	9.5	23.5	20.0	10.0

ArGeMi Cohorts	No personal income	< $50	$51 to 100	$101 to 200	$201 to 300	> $300
MIGRANTS FROM GEORGIA						
Returnees from Moscow	35.0	8.0	11.0	13.5	8.5	24.0
Returnees from other destinations	39.7	5.3	7.3	11.7	8.0	28.0
Potential migrants	46.0	8.0	15.0	20.0	4.0	7.0

Interviewed in Moscow	No personal income	< 4,500 RUB (≈ $152)	4,500-8,000 RUB (≈ $152-269)	8,000-16,000 RUB (≈ $269-539)	16,000-24,000 RUB (≈ $539-808)	24,000-40,000 RUB (≈ $808-1,346)	> 40,000 RUB (≈ $1,346)
	19.5	4.5	8.0	14.5	17.0	20.	13.5

6 Remittances

The volume and impact of monetary flows both on private recipients and on the national economy of sending and receiving countries are much discussed issues not only in literature on international migration, but also in – sometimes speculative - public and political discourses. Since 2004, the following studies have been conducted in the context of migration from the three South Caucasian states: For Armenia, a 2004 study (based on data of 2002), financed by USAID Armenia (Robert & Banaian 2004), followed by a 2005 survey of the *Central Bank of Armenia* on “Assessing the Actual Volumes of Remittances Received by Armenian Households” and a 2006 survey (published in 2008) by the *Asian Development Bank* (ADB) (Tumasyan et al. 2008). While the situation in Armenia was so far researched in country studies only, Georgia was part of a comparative 2006/07 study financed by the *European Bank for Reconstruction and Development* (EBRD) and conducted in four countries (Russia as the most relevant receiving country, Azerbaijan, Georgia and Moldova as sending states). In this chapter, the main results of these surveys will be presented and related to the findings of the 2009 ArGeMi polls.

In the ArGeMi project the definition of remittances followed that of the 2004 USAID financed project on remittances in Armenia: "(...) remittances thus include(s) all funds sent by Diasporan Armenians, whether 'new' or 'old' Diaspora, to households in Armenia without the intervention of a third-party institution making decisions on amount and allocation (use) of funds. Humanitarian assistance is ruled out, because although it is sent to households and supports consumption, it is channeled through state or non-household institutions. In-kind household-to-household transfers should be included, as the recipient household presumably has influence over what is sent" (Robert & Banaian 2004: 3).

As the authors of the above mentioned survey already noticed there exists a difference of the factor 3 in academic literature concerning the volume of remittances to Armenia: "The official estimate of remittance inflows (to Armenia) was $289m in 2003, whereas our alternative estimate equals roughly $900m. Because we examined the official estimate in great detail, we were able to identify where it was most off the mark. The biggest error is due to a methodological mistake that can be rectified at little cost. The true importance of remittances to the Armenian economy is much higher than the ratio of official remittances to GDP (10%) and could be three times as important as that" (Robert & Banaian 2004: 1). For 2009, the World Bank gave the ratio of remittances to the Armenian GDP as 9% (World Bank 2011: 14).

A similar discrepancy of estimates can be observed for Georgia: Whereas the US based firm of Bendixen & Associates which conducted the surveys for the EBRD gives a medium estimate of $418mln of remittances to Georgia in 2006, the *Central Bank of Georgia* named $555mln and the *World Bank* only $393mln (Bendixen 2010: 6).

Based on data from the 2002 Armenian household survey, Robert and Banaian assessed the impact of remittances on social inequality in the following way: "In Armenia, remittances reduce inequality, because the households that receive them would otherwise be at very low levels of income. According to income data reported to the survey, for households receiving remittances, remittances make up 80% of household income on average. Remittances do appear to be going to some of the most vulnerable households in Armenia. We also find that the same percentage of urban and rural households received remittances, but that rural households received relatively more remittances from CIS countries and relatively less from the USA and Canada" (Robert & Banaian 2004: 6).

The 2006 ADB survey draws similar conclusions on the poverty reducing effect of remittances: "Potential, or assumed, absence of remittances would entail high risks of appearing in poverty and extreme poverty for a large share of population, especially in Yerevan and other urban areas. (…) if there were no remittances, it is estimated that extreme poverty among remittance-receiving households would increase 6.9 times, including 15 times in Yerevan. As a consequence, poverty in Armenia would increase by up to 20%, whereas extreme poverty – by 65%. (…) Remittances alleviate the inequality in Armenia. Gini coefficient in Yerevan, under no remittance assumption, would increase from 0.412 to 0.424, while in other urban areas – from 0.367 to 0.392" (Tumasyan et al. 2008: III-IV).

Nearly three quarters of migrants' families in Armenia get remittances from abroad and about 80% of these families receive them regularly. The money is mostly spent on routine expenses and consumption, and only 15% of the receiving families are able to make any savings from the received sums. Indeed, the migrants themselves save money for a number of purposes: 20% for family events, 15% for education of their children and 28% for medical expenses (International Center 2008). Recipients of remittances in Georgia spent about 85% of the money on basic daily expenses such as food, housing, clothing, utilities and medicine (Bendixen 2007: 5). Extra expenses in case of sickness, an accident, physical disability or childbirth can ruin a household, even if the family in question does not belong to the especially vulnerable cohorts of the country. A 2008 Dutch survey graphically summarizes the health care situation:

„In a country where most people struggle to survive for a month, and where any additional cost may lead a family into severe poverty, the need for healthcare is a secondary concern. As from 2006, basic healthcare should be guaranteed by the state and be freely provided, but this remains theory rather than practice. According to surveys undertaken in all regions of Armenia, the lack of access to health care was considered as one of the most difficult problems of vulnerable groups. Health insurances are limited, and especially vulnerable groups have limited access to basic and specialized health care services. Necessary and expensive drugs are often not available at healthcare facilities, and patients have to purchase them on their own. According to surveys, however, groups with privileges, which are often the same as vulnerable groups, are not able in practice to use their privileges with regard to fees and drugs. Corruption is also widespread in the health sector: out-of-pocket payments still constitute an estimated 65 % of all health care expenditure in Armenia. People who refuse paying such 'under the table fees' can expect to get poor treatment, if any treatment at all (The Country of Return Information Project August 2007)" (Johansson 2008: 5).

Characterizing and comparing the pre-crisis patterns of money transfers into the two countries under scrutiny we find as a peculiarity for Armenia that in 2007 the total amount of remittances by Armenians to their country of origin was US$ 846 million, while the average remittances per person were $282 for Armenia, compared with the average remittance for Central and Eastern Europe and the CIS of $114. In the same year the total amount of remittances received in Georgia was $696 million and the per capita amount $158 (UNDP 2009: 160). In international comparison of per capita remittances, Armenia was on the place 27 of remittance-receiving countries in 2007 (UNDP 2009: 160) and on place 26 two years later (World Bank 2011: 14). As a result of the sharp decline of remittances, received in 2010, however, Armenia was no longer listed among the top 30 remittance-receivers in the world (World Bank 2011: 13).

The data on monthly money transfers to Georgia as available from the *National Bank of Georgia* allow the following conclusions: Starting in 2003-2004, the flow of remittances from Russia to Georgia increased at a very high rate, reaching $20-30mln per month in 2005. At the same time the volatility of transfers increased compared to pre-2003. After October 2006, money transfers from Russia to Georgia dropped by $7.3mln per month (Livny et al. 2007: 24).

The EBRD financed surveys on migrant money transfers, conducted both in three countries of origin (Moldova, Georgia, Azerbaijan) and in Russia in late 2006 and early 2007. The surveys revealed that most of the polled migrants in Russia worked in the areas of service, unskilled labor, agriculture or industry and had an average monthly income of €500 (€770 in 2008/9 according to the ArGeMi polls). The authors of the survey, conducted in Russia also noticed that the interviewees had a "somewhat greater formal education than senders we have studied in other countries – in all three groups (i.e. migrants from Moldova, Georgia and Azerbaijan; TS), the majority had completed secondary school or more" (Bendixen 2007a: 1 (146).

The majority of all three cohorts did send money to relatives or friends in their countries of origin in the scrutinized period (2006), in particular in the age group of 35+ years (Bendixen 2007b: 21, 23). However, compared with migrants from Moldova and Azerbaijan, fewer migrants from Georgia transferred money to their friends and relatives (58% against 63% and 64%). Migrants from Georgia had also the lowest share in the group of legal/documented residents (50% against 62% from Moldova and 57% from Azerbaijan), and for this reason obviously had less travelled abroad, but stayed in Russia as full-time residents throughout the entire year (84% from Georgia; 76% from Azerbaijan; 56% from Moldova). Migrants from Georgia also hold the highest shares in the low and medium income groups, with 27% earning an average monthly net income of less than 12,000 RUB (≈ €300) and further 38% earning 12,000-24,000 RUB (≈ €300-600).

Table 14: Remittance senders: Personal net income after paying taxes, 2006

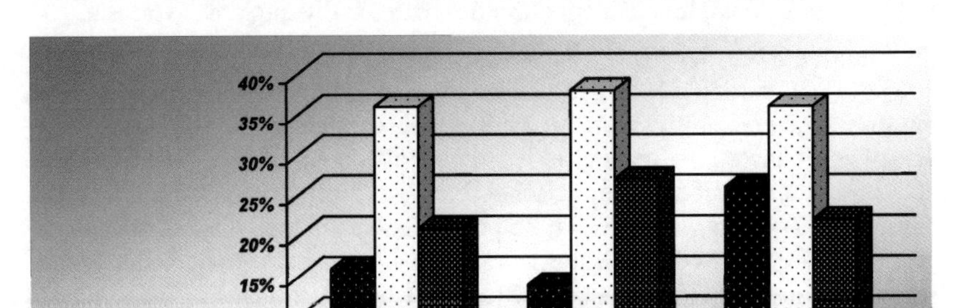

	Azeris	Georgians	Moldovans
More than 24,000 Roubles	16%	14%	26%
12,000 - 24,000 Roubles	36%	38%	36%
Less than 12,000 Roubles	21%	27%	22%

© Bendixen/Associates; EBRD

Corresponding with their relatively low earnings migrants from Georgia transferred an annual per capita average of €750 to their country of origin (migrants from Azerbaijan and Moldova: €850). The total amounts of remittances, sent to the three countries from migrants in Russia are given in Table 15, revealing again the migrant community from Georgia as lowest in money transfer, in particular if compared with the money transfers from Russia to Azerbaijan, which are four times higher:

Table 15: Remittances from Russia to Georgia, Moldova and Azerbaijan

Country	Euros (€)
Poll of Georgian remittance senders in Russia (May 2007)	123 million
Poll of Georgian remittance recipients (January 2007)	140 Million (from Russia only)
Poll of Moldovan remittance senders in Russia (May 2007)	184 million
Poll of Moldovan remittance recipients (January 2007)	181 million (from Russia only)
Poll of Azeri remittance senders in Russia (May 2007)	491 million
Poll of Azeri remittance recipients (January 2007)	310 million (from Russia only

© Bendixen/Associates; EBRD

The EBRD survey conducted in Georgia from December 2006 to February 2007 revealed as a distinctive feature that the profile of remittance recipients was younger than the average adult population. The survey also found that "there (...) seems to be a higher concentration of recipients in the region of Samtskhe Javakheti – mostly populated by ethnic Armenians. In Moldova, remittance recipients are more likely to be older, female, and residents of urban areas" (Bendixen 2007: 4). The province of Samtskhe Javakheti is an economically neglected, poverty stricken area in Southwest Georgia, from where the Armenian majority for lack of employment opportunities traditionally migrates to Russia.

Furthermore the EBRD household poll in Georgia revealed that during 2006 37% of the interviewed respondents had a family member living abroad, but only nine percent of Georgian adults – approximately 317,000 people – received remittances on a regular basis – on the average 8 times a year (Bendixen 2007c: 15). In regional comparison of ten administrative units in Georgia, Samtskhe-Javakheti with 18% of the polled recipients tops the list of most receiving areas, followed by the Western Georgian areas Samegrelo-Zemo Svaneti (12%) and Imereti (11%). Both accommodate a high percentage of IDPs. 45% of all remittances to Georgia come from the migrant communities in Russia, followed by 17% from "Western Europe" (comprising Germany, France, Austria, Belgium, Switzerland, Netherlands, Italy, Spain, Portugal, Finland, England, Norway) and Greece (15%), the USA and "Eastern Europe" (8% each). In Russia, Moscow (54%) stands out as the place from where more than half of all RF remittances to Georgia derive, followed by St Petersburg with only 9%. 37% of the recipients received transfers of $101-200, 33% $100 or less, and 19% received $201-300 (Bendixen 2007c: 22). The average remittance to Georgia in 2006 was $165.

The EBRD surveys also revealed that "a very low percentage of remittance recipients have a bank account and therefore have access to credit and other financial investment products. In Azerbaijan and Moldova, about one fifth of recipients have a bank account, while in Georgia the percentage of those with a bank account is even lower – 11 percent" (Bendixen 2007: 5).

According to the 2006 ABD survey 85% of the remitting migrants reside in Russia and 4.5% in the US, which was the next "largest" host country (Tumasyan et al. 2008: II). The survey also found that the "average size of remittances from Russia is larger than the average size of remittances from almost all other countries" (Tumasyan et al. 2008: II).

The 2009 ArGeMi polls established that migrants from Georgia in Moscow remit more money than those from Armenia. The average monthly per capita amount of money transfers to their homeland is 7,060 RUB, or €168.65 for the sample from Armenia and 7,410 RUB, or €177 for those from Georgia. Related to the official average per capita income of the respondents in Moscow (cf. chapter 5.7), the respondents from Armenia transfer approximately 37% of their average personal income of 7,060 RUB (≈ €461.81) per month, while the financial transfers of respondents from Georgia constitute about 39% of the average monthly personal income (cf. Table 13 for details).

The ArGeMi polls conducted among returnees in Armenia and Georgia are nearly identical. The average monthly per capita amount of money transfers in the four polled samples was ≈ $243 (≈ €178), which makes an approximate growth of 42% in comparison with the amount of $165 (≈ €123.24) as established in 2006/7 by the EBRD poll for recipients in Georgia (cf. Table 16). Comparing the two ArGeMi samples of returnees to Armenia, we find that the returnees from 'other destinations' transfer significantly higher amounts—nearly a third—than those from Moscow. The balance of average monthly per capita amounts of remittance between returnees from Moscow and other destinations is $35.

The ArGeMi data show a considerate national difference in the ratio of transmitting respondents: About three quarters of the respondents in Georgia reported to send money from abroad to relatives and friends, whereas the average percentage in the two samples polled in Armenia is less than two quarters (45.75%). Of these, 39.25% reported to regularly—i.e. monthly—transfer money to Armenia (the average percentage for the samples in Georgia being 77.95%).

Table 16: Money transfers by returnees to Armenia and Georgia according to ArGeMi polls (2009)

Cohorts	Percentage (%) of transmitting respondents	Percentage (%) of regularly transmitting respondents	Total of average monthly amount of money transfer per sample ($) ≈	Average per capita amount of monthly money transfer ($) per sample ≈
MIGRANTS FROM ARMENIA				
Returnees from Moscow	40.5	32.5 (N=81)	15,800	195
Returnees from other destinations	51.0	46 (N=109)	31,850	292
MIGRANTS FROM GEORGIA				
Returnees from Moscow	76.8	78.9 (N=76)	16,300	214
Returnees from other destinations	72.4	77 (N=113)	28,100	249
Average per capita amount of monthly remittances ($) to homeland (all four cohorts (N=379) ≈ $243 (≈ €178)				

The volume margins of money transfers as given in Table 17 include the two samples polled in Moscow and the two samples of potential migrants polled in Armenia and Georgia. The Moscow polls revealed that most migrants transfer small to medium monthly amounts of $134-267 (4,000-8,000 RUB): 40% of the respondents from Armenia, 30.7% of those from Georgia; less than $135 (4,000 RUB): 24.4% from Armenia, 24.8% from Georgia. This is followed by the margin of $267-403 (8,000-12,000 RUB) with 14.4% of the remitters from Armenia and 17.8% from Georgia.

Comparing the means of money transfers, the Moscow polls confirm that the vast majority of transmitters use bank services: 70% among the respondents from Armenia and 62.4% among the respondents from Georgia. The highest share was found among the returnees to Georgia 'from other destinations' (77%), whereas only 27.3% of the returnees 'from other destinations' to Armenia used bank services.

While the above mentioned surveys (including the ArGeMi polls) reflect the pre-crisis situation, the global financial and economic crisis caused a substantial decline of remittances, which has been estimated to be 30-50% for the money transfers from Russia, or a negative balance of $3,690mln for remittances from Russia to all CIS states in 2009 (as compared with 2008), according to the data *Central Bank of Russia* (cf. Table 18). As reasons for the decline the Russian migration expert Irina Ivakhnyuk listed

- The reduction of the number of intra-CIS migrants;
- Wage cuts;
- Devaluation of Russian currency by 25%;
- Increased distrust of the migrants from banks (Ivakhnyuk 2006: 21). Many still remember the complete loss of private savings as a result of the banking crisis in Russia and other former Soviet republics in the early 1990s. Therefore the perspective of another crisis of the Russian and related bank systems may also alter the pattern of money transfers once again.

Table 17: Average monthly amount of remittances and means of financial Transfers

ArGeMi Cohorts	Average monthly amount of remittances sent to homeland from abroad (%)					Means of financial transfer (%)	
	Below $100	From $101 to 200	From $201 to 300	From $301 to 500	Over $500	By bank trans-fer	Cash
MIGRANTS FROM ARMENIA							
Returnees from Moscow	7.0	13.0	9.0	7.0	4.5	30.5	9.0
Returnees from other destinations	1.7	10.7	10.7	8.7	4.7	27.3	9.0
Potential migrants	11.0	17.0	14.0	7.0	-	41.0	9.0
Interviewed in Moscow	Less than 4,000 RUB ($135)	4,000-8,000 RUB ($134-267)	8,000 – 12,000 RUB ($267-403)	12,000-20,000 RUB ($403-672)	More than 20,000 RUB ($672)	70.0	20.0
	24.4	40.0	14.4	5.6	4.4		
MIGRANTS FROM GEORGIA							
Returnees from Moscow	21.1	36.8	23.7	10.5	7.9	71.1	28.9
Returnees from other destinations	20.4	23.9	30.1	13.3	12.4	77.0	23.0
Potential migrants	3.7	13.0	20.4	13.0	5.6	70.4	7.4
Interviewed in Moscow	Less than 4,000 RUB ($135)	4,000-8,000 RUB ($134-267)	8,000 – 12,000 RUB ($267-403)	12,000-20,000 RUB ($403-672)	More than 20,000 RUB ($672)	62.4	25.7
	24.8	30.7	17.8	7.9	5.9		

Despite bleak forecasts for 2009, the year 2010 may have marked certain stabilization: According to the *World Bank* forecasts of late 2010, the volume of remittances from labor migrants to Armenia will reach $824mln in 2010 against $769mln and $1062mln in 2009 and 2008 respectively (WB: Remittances... 2010). The same increase is to be found in the data of the *Russian Central Bank*, according to which the average individual remittance from Russia to CIS countries amounted $548 in October 2010 against $503 in the last quarter of 2009, but remained still below the average $667 throughout 2008 (cf. Table 18).

Table 18: Cross-border remittances via money transfer operators 2003- Oct 2010

	2003	2004	2005	2006	2007	2008	2009	Q1 2010	Q2 2010	Q3 2010	Oct 2010
Total, billions of US dollars											
Money Transfers from Individuals in the Russian Federation	1,310	2,070	3,549	6,005	9,444	13,707	9,967	2,078	2,945	4,013	1,314
to non-CIS countries	...	...	323	622	868	1,098	1,048	352	405	482	164
to CIS countries	...	...	3,226	5,382	8,575	12,609	8,919	1,726	2,540	3,531	1,150
Money Transfers in favor of Individuals to the Russian Federation	588	777	1,041	1,304	1,681	1,977	1,780	413	481	517	177
from non-CIS countries	...	...	645	746	813	796	702	166	191	179	59
from CIS countries	...	...	396	559	868	1,182	1,077	247	290	338	118
Balance[1]	-722	-1,292	-2,507	-4,700	-7,763	-11,729	-8,187	-1,665	-2,464	-3,469	-1,137
with non-CIS countries	...	...	322	123	-55	-302	-345	-187	-215	-303	-105
with CIS countries	...	...	-2,829	-4,824	-7,707	-11,427	-7,842	-1,478	-2,250	-3,193	-1,032
Average remittance, US dollars											
Money Transfers from Individuals in the Russian Federation	...	...	457	536	623	698	513	460	488	561	581
to non-CIS countries	...	...	1,084	1,334	1,407	1,495	1,099	916	918	991	1,008
to CIS countries	...	...	432	510	589	667	483	418	454	530	548
Money Transfers in favor of Individuals to the Russian Federation	...	...	427	481	579	671	588	553	607	703	677
from non-CIS countries	...	...	506	537	583	647	607	606	637	681	673
from CIS countries	...	...	340	425	575	689	576	522	589	716	679

©Central Bank of Russia, http://cbr.ru/eng/statistics/print.aspx?file=CrossBorder/C-b_rem_e.htm

7 Child Separation

Spousal separation and more so the separation of a child from one or both parents represents the most traumatic effect of international migration. However, parental migration and child separation have been very unevenly researched: While immigration, migration and child separation in the American migration system are well studied, the consequences of parallel phenomena in the CIS migration system or the post-Soviet space remain under-researched.

For lack of research, we briefly relate to some of the major findings for the American situation, before addressing to the few studies on the situation of post-

Soviet South Caucasians: The 2002 *Harvard Immigration Project* (Suarez-Orozco et al. 2002) has drawn international attention to the vast number of children, as many as 85% from the Caribbean and Asia, who endure lengthy separation from their parents during the migration process. A 2004 survey among 146 children of Caribbean migrants found that "children separated from parents because of migration were more than twice as likely as other children to have emotional problems although their economic status was improved. One-third had serious levels of depression or interpersonal difficulties affecting schooling and leading in some cases to suicidal ideation. Differences were found in relation to gender & ethnicity" (Jones et al. 2004).

The high-school drop out rate was 40.7% among all children from immigrants to the US, who had been separated from their students during migration, compared with only 12.6% of those who migrated together with their parents (Gindling & Poggio 2009: 3, Table 3). Even if children are left in the care of relatives or friends, the separation from a parent may cause psychological damage that is manifested in deviant, "acting out" behavior or poor self esteem and depression.[14] For the parents, parental migration is likewise traumatic and often manifests in sadness, guilt and anxiety over the separation (Suarez-Orozco et al. 2002).

Family reunion in adolescence may cause additional problems, „when children are battling with developmental issues of identity and figuring out where they belong. Also, they are likely once again to experience migratory separation when they separate from the surrogate parent and friends they had grown accustomed to over the years. Upon arrival in the host country these children may face reconstituted families with step-parents and siblings and have to figure out how to fit in with this new family. Sometimes the parent is working more than one job or working and going to school, leaving little or no time to help them settle in their new environment, nor for adequate supervision. Recent immigrant children may also have to struggle with differences in language, accent, social systems as well as race classification. In addition, many encounter selection procedures at school and a school environment that considers them to be at a disadvantage to their (...) peers (…). When both parents' and children's expectations about the reunion are not met, the child may react with anger and rebel and parents may label this behavior as ingratitude and resort to harsh methods of discipline" (Pottinger & Williams Brown 2006).

In the ArGeMi cohorts the relative or even absolute majorities of respondents have no children, partly due to the respondents' young age, as it is the case

---

14 Cf. Glasgow, G.F.; Ghouse-Shees, J.: 'Themes of rejection and abandonment in group work with Caribbean adolescents'. *Social Work With Groups*, 1995, 17, 3-27; Pottinger, Audrey M.: ‚Disrupted caregiving relationships and emotional well-being in school age children living in inner city communities'. *Caribbean Childhoods: From Research to Action*, 2005, 2, pp. 38-57; Jones et al. 2004

among the potential migrants in Georgia (66%). The equivalent cohort in Armenia shows opposite results: Only 47% of the potential migrants had no children yet, whereas 33% - the highest ratio of that margin in all Armenian samples – reported one child. Returnees from other destinations than Moscow and Russia have the most children both in the Armenian and Georgian samples. Among the interviewees in Moscow more than every fourth migrant from Armenia has a child left in his or her country of origin, while more than every tenth migrant has left two children. The shares are less for interviewees from Georgia: Nearly every fourth has a child in Georgia, while 22% reported to have one child in their Muscovite household (interviewees from Armenia: 27.5%). In comparison, respondents from Georgia tend to have their children with them in Moscow, whereas more respondents from Armenia reported to have their children left in their country of origin.

Table 19: Number of children (younger than 16 years) in household of ArGeMi respondents (%)

ArGeMi Cohorts	No child	1 child	2 children	3 children
Migrants from Armenia				
Returnees from Moscow (interviewed in Armenia)	55.0	22.5	19.5	3.0
Returnees from all other destinations	48.3	25.7	24.7	1.3
Potential Migrants	47.0	33.0	17.0	2.0
In Moscow				
Household in Armenia	51.1	27.8	11.1	4.4
Household in Moscow	63.0	27.5	9.0	0.5
Migrants from Georgia				
Returnees from Moscow (interviewed in Georgia)	56.0	25.0	16.5	2.5
Returnees from all other destinations	52.0	28.0	16.3	3.7
Potential Migrants	66.0	24.0	8.0	2.0
In Moscow				
Household in Georgia	25.3	19.3	6.0	2.4
Household in Moscow	66.5	22.0	11.0	0.5

The ArGeMi respondents were also asked who the caregiver of their child was during their last stay abroad. With regard to **transnational parenthood** the survey revealed that less than a fifth of all returnees from Moscow to Armenia and Georgia had been accompanied by a child or children. The percentage among the returnees from other destinations was even lower, in particular among the returnees to Georgia (7.3%), while 17.5% of the returnees to Armenia had travelled accompanied by the child/children. The cohorts, interviewed in Moscow reported higher percentages of personal care giving (27.5% of the migrants from Armenia, 34% of those from Georgia). However, the absolute majorities of the two Moscow cohorts admitted to have found 'other' – unspecified – solutions for the child (66% of the migrants from Armenia and 57.6% of those from Georgia). This 'other solution' seems not to imply that their children were transferred to an orphanage or similar institution in the homeland, because there was a special margin for this suggestion in the questionnaire. It is quite noteworthy that with one exception in Moscow none of the respondents in all other cohorts admitted to have given his/her child to an orphanage, although this phenomenon has been critically disputed at least in Armenian media of previous years. We suspect the interview results to be triggered by the bad public reputation that such institutions have, and second by the wide-spread stereotype of the selfless Armenian/Georgian mother.

Presumably, the answers of the interviewees have been influenced by cultural traditions, public discourse and gendered narratives. In Armenia, the traditional gender division of labor seems particularly widespread, with a migrating breadwinner and the mother remaining in the country with the children. If a female child caregiver, in particular the biological mother, migrates, her choice must be justified according to societal expectations. The selfless or even self-sacrificing mother is such a traditional role model that allows compatibility with female migration, even if the children under the woman's/mother's care remain separated from her for a long time (Ishkhanian 2002 and 2003). In this case, in the general societal perception emotional losses seem compensated by material gains for the child/children and other family members.

Table 20: Who took care for respondent's child during his/her recent stay abroad?

ArGeMi Cohorts	The child/children was under my care	Spouse/ relatives/ friends took care	The child/children was in an orphanage in homeland	Other solution
MIGRANTS FROM ARMENIA				
Returnees from Moscow (interviewed in Armenia)	19.0	41.5	-	-
Returnees from all other destinations	17.5	50.0	-	-
Interviewees in Moscow	27.5	6.0	-	66.0
MIGRANTS FROM GEORGIA				
Returnees from Moscow (interviewed in Georgia)	19.5	30.5	-	-
Returnees from all other destinations	7.3	31.3	-	-
Interviewees in Moscow	34	7.5	1.0	57.6

Biological parenthood and social, financial and/or educational care giving are obviously not always identical, and the latter is more significant at least according to the findings of the ArGeMi project. Although the relative or even absolute majorities of the ArGeMi migrant cohorts do not have own children, in three of the four cohorts of returnees the absolute majorities do give care to a child - on the average even two children: 61.7% of all returnees to Armenia from other destinations, 59.5% of the returnees from Moscow to Georgia and 51.3% of returnees to Georgia from other destinations. The highest single mention comes from the margin with two children in the cohort of returnees to Armenia from other destinations (38%). The ratio of caregivers among the interviewees in Moscow was relatively low: 36% of all migrants from Armenia and 42.5% of those from Georgia give care to one and more children:

Table 21: Number of children (younger than 16 years) the respondent takes for (financially and/or as child care giver)

ArGeMi Cohorts	No children	1 child	2 children	3 children	More than 3 children
Migrants from Armenia					
Returnees from Moscow (interviewed in Armenia)	36.5	16.5	32.0	12.5	2.5
Returnees from all other destinations	28.3	18.0	38.0	14.0	1.7
Potential Migrants	28.0	13.0	37.0	19.0	3.0
Interviewees in Moscow	56.0	16.5	15.5	3.5	0.5
Migrants from Georgia					
Returnees from Moscow (interviewed in Georgia)	40.5	24.0	31.5	4.0	-
Returnees from all other destinations	48.7	23.0	24.0	4.3	-
Potential Migrants	66.0	15.0	15.0	4.0	-
Interviewees in Moscow	53.5	18.5	16.5	7.0	0.5

8 General Evaluation of Migration

The respondents of the ArGeMi survey showed a clear understanding of the dual and even conflictual nature of international migration that may enhance, on the one hand, personal liberties, employment and income opportunities, while on the other hand negative individual and national effects become increasingly noticeable in the countries of origin: spousal and child separation, changes in the demographic composition with possible feminization, increased and accelerated aging of the affected societies, decline of taxes, revenues etc. In the ArGeMi questionnaire, these individual and collective pros and contras of migration were highlighted in four corresponding questions. The perception of migration as ambivalent indirectly showed in the fact that 13-32% of the respondents were uncertain how to answer to these positive or critical assumptions.

On the whole, the respondents from Armenia displayed a more critical attitude than the respondents in the four samples from Georgia: Roughly every third in the Georgian cohorts 'fully agreed' with the statement that 'migration is a blessing for Georgian people, because they can freely travel and work abroad' (cf. Table 22), while about every second respondent in all samples from and in Armenia fully disagreed with this assumption. Somehow unexpectedly, with 53.3% the critical attitude toward out-migration was most explicit among the sample of migrants from Armenia to other destinations than Moscow, while the ratio in the Georgian equivalent was only 18.3%.

Table 22: Curse or blessing? Evaluation of migration effects on the nation and country of origin (%)

ArGeMi cohorts	Do you believe that out-migration is a blessing for Armenian/Georgian people – they can travel and work abroad freely?					Do you believe that out-migration is a blessing for Armenia/Georgia – it relieves the labor market and brings money into the country?				
	Fully agree	Partly agree	Partly disa-gree	Fully disa-gree	Not sure	Fully agree	Partly agree	Partly disa-gree	Fully disa-gree	Not sure
MIGRANTS FROM ARMENIA										
To Moscow (interviewed after return to Armenia)	12.0	34.0	-	49.5	4.5	11.5	45.0	-	41.0	2.5
To other destinations	11.3	32.3	-	53.3	3.0	11.0	39.7	-	47.3	2.0
In Moscow	27.5	39.5	8.5	6.0	18.5	7.0	28.5	13.5	18.5	32.5
Potential Migrants	12.0	18.0	20.0	46.0	4.0	7.0	22.0	20.0	46.0	5.0
MIGRANTS FROM GEORGIA										
To Moscow (interviewed after return to Georgia)	32.0	44.5	-	17.0	6.5	32.5	49.0	-	10.0	8.5
To other destinations	31.3	45.3	-	18.3	5.0	31.7	49.7	-	13.7	5.0
In Moscow	29.5	43.5	11.0	3.5	13.0	9.0	35.0	20.5	13.5	22.0
Potential Migrants	38.0	37.0	11.0	4.0	10.0	37.0	40.0	8.0	5.0	10.0

In particular the issue of brain waste has touched a nerve among migrants from Armenia (cf. Table 23): 60.7% of the returnees from other destinations “fully agreed” that “migration is a curse, for the country looses its best people”, while another 26.7% from the same sample expressed partial agreement with the statement. Skepticism about out-migration had the lowest intensity in the sample of returnees to Georgia from other destinations than Moscow.

Table 23: Curse or blessing? Evaluation of migration effects on nation and country of origin (%)

ArGeMi Cohorts	Do you believe that out-migration is a curse for Armenian/Georgian people – migrants suffer exploitation and discrimination abroad, families suffer as well?					Do you believe that out-migration is a curse for Armenia/Georgia – the country loses its best people?				
	Fully agree	Partly agree	Partly disagree	Fully disagree	Not sure	Fully agree	Partly agree	Partly disagree	Fully disagree	Not sure
MIGRANTS FROM ARMENIA										
Returnees from Moscow	27.0	51.5	-	19.0	2.5	41.0	41.0	-	17.0	1.0
To other destinations	37.7	39.3	-	19.3	3.7	60.7	26.7	-	10.0	2.7
In Moscow	12.0	27.0	14.5	20.5	26.0	35.0	27.0	14.5	7.0	16.5
Potential migrants	38.0	36.0	13.0	9.0	4.0	47.0	38.0	7.0	4.0	4.0
MIGRANTS FROM GEORGIA										
Returnees from Moscow	18.5	50.0	-	18.5	13.0	17.0	43.0	-	11.0	9.0
To other destinations	15.7	49.3	-	29.7	9.3	34.3	44.3	-	13.7	7.7
In Moscow	10.0	28.5	21.5	15.0	25.0	26.5	36.0	11.5	10.0	16.0
Potential migrants	12.0	39.0	13.0	18.0	18.0	30.0	36.0	9.0	11.0	14.0

We presume that the divergence in the general perception of migration stems from the collective self-image of Georgians and Armenians and the history of migration in both countries: Although in 16th century both nations suffered massive compulsory deportation and re-location by Iranian invaders, such painful events influenced the Armenian thinking more profoundly than the Georgian one. In the Armenian case the forcible population transfer from Nakhichevan, the Ararat plain and the Van province came as just another episode in a seemingly endless chain of expulsion and massive flight that centuries of power struggle over the Armenian Highland had forced upon its population. Post-Soviet out-migration is seen as part of this experience and not as the exception from a general rule of being repeatedly forced out of the homeland. In the Georgian case, several centuries of glorious past (10th to early 13th centuries), when a centralized, strong Georgian kingdom emerged support a more positive view on national history and prevent that the experience of massive involuntary emigration in modern times dominates the Georgian narrative of national history.

9 Personal Migration Experiences: Contentment and Insecurity

The personal experiences of the ArGeMi respondents reveal a) discernment similar to the differentiated general evaluation of migration, and b) a curious discrepancy between the skeptical migration perceptions as revealed in particular by the respondents from and in Armenia and their personal contentment with migration. In other words: Migrants from Armenia are generally skeptical about migration, but personally happy to live abroad or to have migrated. On the average, migrants from Georgia feel and behave exactly the opposite.

For example, the respondents' skeptical evaluation of 'brain waste' only partly corresponds with own experience, for only 18.7% of the returnees from to other destinations than Moscow mentioned that their occupation abroad did 'not at all' correspond with the level of their education. However, the ratio of discontents is higher among interviewees in Moscow: About every fourth migrant from Armenia and Georgia there mentioned the qualitative discrepancy of his/her education and the requirements of the job. As seen from another point of view, nearly 70% of the returnees to Georgia from other destinations than Moscow admitted that the level of their education did 'not at all' correspond with their last employment abroad. But only a third of the same sample – i.e. half of the ratio among respondents from Armenia - concluded that migration 'is a curse because the country looses its best people'.

As mentioned before, migrants from the post-Soviet space possess a rather high level of education (cf. Table 10), which makes the frequent experience of '3 D jobs' in labor migration – dirty, degrading and dangerous – especially painful or humiliating. Low skill employment is often combined with little security and low wages.[15] But surprisingly enough, low skill employment obviously did not affect the satisfaction of the ArGeMi respondents, for more than every second returnee to Armenia was 'entirely' satisfied with his or her last stay abroad. Here, the feeling of satisfaction is obviously determined by other criteria than employment conditions or the correspondence of the personal education with employment levels. The most satisfied sample were the returnees to Armenia from other destinations than Russia (66.7%), whereas the equivalent sample in Georgia was nearly half less content (34.6%). Furthermore, the highest poll of 'entire satisfaction' (66.7%) has been found in the sample of returnees from other destinations to Armenia, which at the same time showed the highest agreement (60.7%) with the assumption that Armenia looses her best people from migration!

In general, the mention of complete satisfaction was lower among the samples in and from Georgia, with the exception of those, who were interviewed in Moscow (42% against 40.5% of the respondents from Armenia). It is also noteworthy that the margin with the highest percentage of 'entirely satisfied' respondents from Georgia is to be found among the returnees from Moscow, whereas the lowest degree of 'entire satisfaction' was to be found among the returnees from other destinations to Georgia. In general, all interviewees in Moscow expressed lesser contentment than the returnees interviewed in their

15 Cf. „Amnesty International", Web-Action, 12 September 2005. http://amnesty.name/en/library/asset/ACT30/027/2005/en/ca1dd0b3-d470-11dd-8743-d305bea2b2c7/act300272005en.pdf

countries of origin. This may be explained by the fact that interview situations during or after a migration cycle are psychologically quite different.

Table 24: Are you generally satisfied with your (last) stay abroad? (%)

Cohorts of ArGeMi Respondents	Entirely	Partly	Not at all
MIGRANTS FROM ARMENIA			
Interviewed in Moscow	40.5	54.0	5.5
Returnees from Moscow (interviewed in Armenia)	52.5	40.0	7.5
Returnees from other destinations	66.7	29.7	3.6
MIGRANTS FROM GEORGIA			
Interviewed in Moscow	42.0	53.0	5.0
Returnees from Moscow (interviewed in Georgia)	46.5	45.5	8.0
Returnees from other destinations	34.6	48.7	16.7

Another striking contrast to the overall personal contentment with the last migration trip consists in the feeling of insecurity, which was experienced by more than every fourth returnee from Moscow and more than every tenth returnee from other destinations than Russia. Moreover, every third of the interviewees in Moscow reported that he or she had been offended in that city, while 15.5% of the returnees to Armenia experienced threat or menace when being Moscow. Finally, in the Moscow samples 13% of the respondents from Armenia and 10.5% of those from Georgia have suffered physical attacks or intentional injuries. While in the Moscow samples offence concerned the respondents from Georgia more than those from Armenia – a difference of 6% -, threat or menace has been reported nearly twice more among the returnees from Moscow in Armenia (15.5%) than among the returnees to Georgia (8%). Among those, who suffered assaults and injuries, the difference between respondents from Armenia and Georgia in Moscow was insignificant (2.5%). More than every tenth Armnenian returnee from other destinations than Moscow experienced offence, while in the equivalent Georgian sample the ratio was only 4.3%.

Table 25: Experience of insecurity, verbal and physical violence

Cohorts of ArGe-Mi respondents	„During my last stay abroad I felt insecure...“	„While abroad, I experienced...“		
		...Offence	...Threatened or menaced	...Physically attacked/intentionally injured
Migrants from Armenia				
Interviewed in Moscow	11.0%	32.0%	4.5%	13.0%
Interviewed in Armenia after return from Moscow	26.0%	19.5%	15.5%	3.5%
Interviewed in Armenia after return from other destinations	10.7%	13%	1.0%	3.3%
Migrants from Georgia				
Interviewed in Moscow	26.0%	38.0%	9.0%	10.5%
Interviewed in Georgia after return from Moscow	26.0%	18.5%	8.0%	1.0%
Interviewed in Georgia after return from other destinations	11.3%	4.3%	3.0%	1.0%

How can negative emotions such as insecurity or experiences such as offence and assault be consistent with general migration contentment? A weak tradition of human and in particular worker rights standards and a corresponding unawareness of such rights and their violations may provide the answer.

10 Individual Self-Reliance in Absence of Legal Rights

Closely linked with the above mentioned absence or weakness of legal rights is the phenomenon of individual self-reliance: Regional migrants in the post-Soviet space do not count on national legislation, law or law enforcement, neither in their homeland, nor in their host lands. In the course of the ArGeMi project a country comparison of migration legislation revealed that this sphere of legislation is still incomplete, although a number of migration-related agreements on a bilateral Armenian-Russian level already exist. However, the nearly complete lack of any Russian-Georgian regulations of migration is striking. Migration between both countries takes place in a legal vacuum.

On international level, neither Russia, nor Armenia and Georgia have signed or ratified the United Nations' *International Convention on the Protection of the Rights of All Migrant Workers and Members of their Families* (1990; in force 1

July 2003). Azerbaijan is the only of the three South Caucasian states that has signed and ratified (in 1999) this important safeguard of the migrants' human rights.[16] Internationally there are only 33 signatories and 44 parties to this UN convention. Of the four *ILO Conventions* on migrant rights that have been issued since 1919 neither the USSR, nor the Russian Federation ratified any. Among the South Caucasian successors of the USSR only Armenia has ratified in 2006 the *Migration for Employment Convention* in its revised version of 1949, together with the 1975 *Migrant Workers (Supplementary Provisions) Convention* (C 143).[17]

The lack of reliable safeguards, or of substantiated legal regulations for the migration between sending and receiving countries and the inefficient law enforcement may be regarded as the main reason for the self-reliance displayed in the interviews. Self-reliance is the individual response to failing states, administrations, institutions or justice. Correspondingly, in a case of emergency the absolute majority of all respondents in Moscow - 52% of the respondents from Armenia and 55% of those from Georgia - contact their local friends or acquaintances for help and advice. Some 38% of the migrants from Armenia and 46.5% from Georgia would seek help rather from compatriots or expats. Only negligible minorities would address to their diplomatic or church representations or would seek help from third non-governmental organizations. Among these few votes, the 'highest' trust into the diplomatic representations was displayed by the two cohorts of potential migrants: 13% of potential migrants in Armenia and 8% of those in Georgia would address to their embassies in emergencies abroad.

The ArGeMi cohorts which had been interviewed after return to their countries of origin show modified, but not generally different patterns of behavior. The Armenian cohorts in general revealed a significant tendency to seek help and advice less on an ethnic basis, but in the first place from their local acquaintances no matter what their ethnicity/nationality may be: 32.5% of the returnees from Moscow and 25.3% of the returnees from other destinations, who had been interviewed in Armenia, do so. Even among the cohort of potential migrants the share of those who prefer to seek help from acquaintances rather than expats and compatriots is slightly higher (25% against 22%). The cohorts interviewed in Georgia displayed a somehow more 'patriotic' attitude, giving the preference to compatriots rather than to 'local people you know': 46% of all interviewed potential migrants in Georgia, 44.5% of the returnees from Moscow and 38.7% of the returnees from all other destinations rather seek help from their compatriots/expats (the percentages for the 'local people' margin being: 29% - 42.5% - 32.7%).

---

16 Nor are Germany or the US signatories. Cf. http://treaties.un.org/Pages/ViewDetails.aspx?src=TREATY&mtdsg_no=IV-13&chapter=4&lang=en

17 Cf. http://www.ilo.org/ilolex/english/subjlst.htm

Every tenth Georgian returnee from Moscow and every 13th returnee from other destinations do not even expect any help or advice abroad:

Table 26: In case of an emergency - where do you seek help and advice when in trouble in Moscow/abroad? (%)

	Migrants from Armenia				Migrants from Georgia			
Help/ advice are sought from:	Inter-viewed in Moscow	Return-ees from Moscow	Return-ees from other destina-tions	Potential migrants	Inter-viewed in Moscow	Return-ees from Moscow	Return-ees from other destina-tions	Potential migrants
Compatri-ots	38.0	18:=	13.3	22.0	46.5	44.5	38.7	46.0
Local people you know	52.0	32.5	25.3	25.0	55.5	42.5	32.7	29.0
Armenian /Georgian Embassy	3.5	1.5	2.7	13.0	10.0	2.0	8.3	8.0
Local authorities	15.0	1.5	1.7	9.0	14.5	0.5	4.0	-
Armenian / Georgian NGOs	3.5	2.0	4.0	4.0	2.5	0.5	1.0	-
Armenian / Georgian church represent-atives	1.0	-	1.0	4.0	2.0	-	-	-
Human and civic rights NGOs	5.0	2.5	2.0	4.0	2.0	-	0.3	-
Other organiza-tions	2.0	1.5	1.7	-	5.5	-	0.3	-
Lawyer	7.5	1.0	0.3	-	8.5	-	1.7	-
Others	6.5	1.0	1.0	-	5.5	-	-	-
One cannot expect any help in Moscow / abroad	3.5	4.5	3.3	2.0	6.0	10.0	13.0	8.0
No answer	12.0	34.0	43.7	15.0	2.5	-	-	9.0

The general distrust of authorities and institutions that the above quoted ArGeMi polls reveal must be interpreted as a consequence of a more general lack of social cohesion in post-Soviet societies. However, as a comparative examination of *Gallup* data for the two South Caucasian countries Armenia and Azerbaijan showed, there exist significant differences in this region: "Armenians and Azerbaijanis were asked how they would describe the loyalty they felt toward their country. Among all CIS countries, Azeris and Tajikistanis appeared the most resolutely loyal—about two-thirds of each population selected the following response: 'my loyalty is unshakable—it's my country whether it is on the right development path or not.' Among Armenians, just 42% selected this response.

Similarly, when the two populations were asked how strongly they identified with their country (…) the percentage responding 'strongly' or 'very strongly' was much higher in Azerbaijan (75%) than in Armenia (48%)" (Gradirovski & Esipova 2007: 3).

In the same poll Armenians revealed a low percentage of confidence to their government ('doing an excellent' or 'good job': 42% in Azerbaijan, 17% in Armenia) and in the economic perspectives (74% being optimistic in Azerbaijan, only 47% in Armenia), "which also influences on the respondents' determination to stay in the country. Being asked whether they would move permanently to another country if they had the means to do so, about three-fourths of Azerbaijanis (74%) said they would stay in Azerbaijan, while only 55% of Armenians said they would stay in Armenia. This closely corresponds with the results for the question 'Do most children in your country have the opportunity to learn and grow, or not?' In Azerbaijan two-thirds (67%) answered 'yes' (this percentage is below only Belarus's 76%), whereas in Armenia just under half, 48%, answered 'yes'. It is hard to underestimate the importance in any culture of perceptions regarding the treatment of children" (Gradirovski & Esipova 2007: 4).

The advanced individualization among Armenian nationals that the above quote polls likewise reveal puts also the widespread assumption of influential Diasporic networks into question. It is generally believed that in particular Armenians who look back on a millennium of expulsions, relocations, massive flight and migration developed closely knit networks that may facilitate and even trigger further emigration. However, reality seems more complicated and complex: First, in the ArGeMi polls migrants from Armenia showed a clear tendency to relay on personal contacts rather then ethnic networks. Second – and according the author's personal experiences with Armenian Diasporas in numerous European and Near East countries, - the stratification inside these communities is highly advanced. In Russia, there exists no monolithic Armenian community, but rather a pluralism of several communities centered by principle of origin: Armenians who originate in Diasporic communities in the South Caucasus (Georgia, Azerbaijan), Central Asia or the European parts of the previous USSR generally feel that they have more in common with other Armenians from those countries than with Armenians from Armenia. The 'expat factor' or the country of origin is more determining than the ethnic affiliation.

11 Return and Re-Migration Potentials

On the average, experts and stakeholders assess the chance of return of migrants as very unlikely. In all eight ArGeMi cohorts the majority of respondents did not believe that 'most migrants from Armenia/Georgia will return to their countries of origin'. The samples from Armenia reveal higher amplitude, because they comprise both the highest skepticism against a return and the highest optimism in return: Only 24.7% of the returnees from other destinations expect a return, while 72.5% of the Armenian interviewees in Moscow doubt the return (66% of the Georgian equivalent sample).

On the other hand, 44% of the potential migrants in Armenia believe in the return of most migrants, which was the most optimistic vote of all eight cohorts (closely followed by the Georgian equivalent sample with 39%). As a general trend the two cohorts of potential migrants displayed the highest confidence in a return, whereas the interviewees in Moscow – probably judging by their own examples – were the most skeptical. In comparison of the Armenian and Georgian situations, differences in trends can be perhaps explained with the higher significance that migration related issues have in the traditional thinking. The idea that all Armenians should live in their own national state dominates Armenian political thinking since the turn of the 19th to the 20th century. But massive emigration runs counter to this traditional ideal and is therefore more confusing and difficult to accept than in Georgia.

Table 27: General return assessment – "Do you believe that most migrants from Armenia / Georgia will return to their countries of origin"?

ArGeMi Cohorts	Yes (%)	No (%)	Difficult to say
MIGRANTS FROM ARMENIA			
Returnees from Moscow (interviewed in Armenia)	35.5	48.5	16.0
Returnees from all other destinations	24.7	55.0	20.3
Interviewees in Moscow	27.5	72.5	-
Potential migrants	44.0	42.0	14.0
MIGRANTS FROM GEORGIA			
Returnees from Moscow (interviewed in Georgia)	36.5	28.0	35.5
Returnees from all other destinations	33.3	23.0	43.7
Interviewees in Moscow	34.0	66.0	-
Potential migrants	39.0	12.0	49.0

Asked about their personal return intentions, the two cohorts, 27.5% of the respondents from Armenia and even 40.5% of those from Georgia declared to return to their countries of origin in the course of the next six months. However, the vast majority will only temporarily return: 69.1% of the respondents from Armenia and 59.3% of those from Georgia come back for holidays or vacations. The next highest percentage is found in the mention of involuntary return: 28.4% of the respondents from Georgia and 23.6% in the equivalent Armenian sample expect to be deported by the RF authorities, which is a surprisingly high percentage for the cohort from Armenia, considering the fact that this cohort does not face similar problems to enter Russia and stay there legally as migrants from Georgia do. The figure may be also indicative for a still high percentage of undocumented nationals from Armenia. Expert estimates on the ratio of illegal labor migration from Armenia to Russia are as high as 90%.[18]

The mentions of all other possibilities offered in the ArGeMi questionnaire – expiration of visa, completion of studies, or termination of a work contract – were insignificant, compared to the two major reasons of temporary voluntary return and compulsory deportation. The results among the cohorts of returnees interviewed in their countries of origin are rather diversified: 63% (64% of the equivalent Georgian sample) of the returnees from Moscow to Armenia named personal reasons for their return, followed by 19.5% who returned after their work or studies in Moscow had been completed (only 7.5% in the Georgian sample). The sample of returnees from other destinations named even more diversified reasons: 35% mentioned the expiration of visa (only 18.3% in the Georgian equivalent sample), 30.7% personal reasons (46.3%), and nearly a quarter the completion or termination of work and studies (29.7%). Only 4% in the two samples of returnees from other destinations admitted to have been deported by the authorities of their host countries. Compared with the high percentages of 23.6% and 28.4% of respondents in the two Muscovite samples, who expect to be deported from Russia this is a rather unlikely result. It may be explained by the differences in the interview situation: While the cohorts of returnees and potential migrants were interviewed in their countries of origin and by compatriots, the Muscovite samples were interviewed by representatives of the Russian titular nation/majority. 'Failures' or unsuccessful migration is easier to admit to strangers than to compatriots.

Most returnees who had been asked in Armenia and Georgia about possible plans of remigration denied such intentions for the next twelve months; for obvious practical reasons, the denial was highest among the returnees from Moscow to Georgia (60.5%), followed by 51.5% of the equivalent sample in Armenia. Re-migration intentions were highest (44.3%) among the returnees from other destinations to Armenia. A fifth of these planned short term stays abroad (1-3 month), whereas the majority of the re-migration interested respondents in Georgia planned medium-termed stays (3-6 months): 51.4% of the returnees from Moscow and 41% of those from other destinations. As main purposes for their intended next stay abroad most respondents mentioned employment, followed by visits to relatives and education.

---

18 According to the *Eurasia Partnership Foundation* (EPF), related to the situation after the RF amendments on immigration legislation in 2007. – Cf. http://epfound.am/files/labor_migration_eng.pdf

Table 28: Respondents' (re-)migraton intentions

'Do you plan to stay abroad longer than one month during the next twelve months?'	Yes (%)	No (%)	Difficult To say (%)	If Yes: Intended duration of stay abroad (%)		
				1-3 months	3-6 months	More than 6 months
MIGRANTS FROM ARMENIA						
Returnees from Moscow (interviewed in Armenia)	30.5	51.5	18.0	8.5	5.0	16.5
Returnees from all other destinations	44.3	46.0	9.7	20.7	10.7	15.0
MIGRANTS FROM GEORGIA						
Returnees from Moscow (interviewed in Georgia)	18.5	60.5	21.0	16.2	51.4	-
Returnees from all other destinations	35.0	47.3	17.7	20.0	41.0	

The results given in Table 28 coincide with those from a 2009 *Gallup* poll in 12 post-Soviet republics, which revealed that the emigration/migration potential was very high: In addition to existing migration, an overall amount of estimated 70 million people would like to migrate temporarily and 30 million permanently. In this survey, Armenia came second after Moldova (54%) with the intention of more than one third of its population (39%) to permanently leave the country; the same percentage of respondents in Armenia wanted to study or participate in a work-study program abroad – the highest amount in all polled countries – while 44 percent would like to move for temporary work. The comparison with polls from Armenia's two South Caucasian neighbor states reveals a thrice lower result of emigration intentions: Only 14% of the respondents from Georgia and 12% of those from Azerbaijan desired to move to another country permanently (Barsoumian 2010).[19]

Armenian readers' reactions to the publication of the Gallup polls were highly emotional. In an online comment one reader wrote, much to the consent of others: "I wonder who sponsored this poll. The Diaspora already knew that many would like to leave Armenia. If this poll is even roughly close to the reality, it is psychologically devastating, and it gives our enemies more encouragement to suffocate Armenia. I hope that this poll will shock the authorities to work on improving the social conditions of the people and to establish a lawful society. The government should create a welcoming atmosphere to attract diasporans to settle to their fatherland and increase its population. But the government has done the opposite by allowing criminal elements to control the country and demoralize the people to a point of depopulating the fatherland" (Barsoumian 2010).

The USA and Russia that in general represent the two main attractive international destinations for regional migrants from the South Caucasus are also the two most frequently named destinations for those ArGeMi respondents who intend to migrate in near future. However, there are significant differences in the country comparison as well as in the comparison of destination choices for work

19 The poll in Armenia was based on interviews with 1,000 respondents.

and residence. Provided that they were given the opportunity to freely choose a country of entry not regarding any visa or other regulations, two of the four cohorts from Armenia named the USA as the most favorable destination for work, with the exception of the returnees from Moscow and the sample of potential migrants. A quarter of the potential migrants from Armenia would still prefer Moscow as a destination for employment abroad, while 22% voted for the USA, thus displaying a near-to equivocal destination preference. The equivalent sample from Georgia reveals a reverse and unequivocal trend: While only 2% of them prefer Moscow, 26% voted for the USA. At the same time the poll of this sample, together with the sample of returnees from other destinations, clearly indicates the increasing 'Europeanization' in the Georgian migrants' destination choices, with Italy (18%), Great Britain (13%), Spain (8%) and Germany (7% of the potential and even 12.7% of the returnees 'from other destinations') as the most frequently named European destinations. In comparison the polls from the cohorts in and from Armenia lack this trend, showing more diffuse results.

Table 29: Favorite migration destination for work (%)[20]

	Migrants from Armenia				Migrants from Georgia			
Country of destination	Interviewed in Moscow	Returnees from Moscow	Returnees from other destinations	Potential migrants	Interviewed in Moscow	Returnees from Moscow	Returnees from other destinations	Potential migrants
Russia	11.0	14.0	10.7	25.0	15.5	21.0	1.3	2.0
USA	18.5	7.5	22.0	22.0	15.5	7.0	20.7	26.0
France		2.5	6.7	9.0		2.5	3.7	6.0
Germany	4.0	2.0	7.3	7.0	4.5	3.5	12.7	7.0
Greece		1.0	1.0	3.0		-	4.0	3.0
Italy		0.5	3.7	4.0		4.5	4.0	18.0
Spain		1.0	2.7	5.0		2.0	3.7	8.0
Netherlands		-	3.0	5.0		-	0.7	-
Great Britain	3.0	1.5	1.7	3.0	1.0	4.5	8.3	13.0

As it was expected by the designers of the ArGeMi questionnaire most interviewees distinguished between a migration destination for work and one for residence: A country that is perceived as a good choice for employment, business or studies must not necessarily be ideal for living there. This divergence becomes obvious when comparing the polls for the USA or Germany as countries for work or residence. In particular the samples of returnees from other destinations comprise lower votes to both states as countries for residence. The polls of the two samples of potential migrants in Georgia and Armenia show their preference of France and Southern European destinations, while potential migrants in Georgia as only cohort 'discovered' Great Britain as a favorable destination choice both for work (13%) and for residence (11%):

20 Selection of the nine most frequent mentions

Table 30: Favorite migration destination for residence

	Migrants from Armenia				Migrants from Georgia			
Country of destination	Inter-viewed in Mos-cow	Return-ees from Moscow	Return-ees from other destina-tions	Potential migrants	Inter-viewed in Mos-cow	Return-ees from Moscow	Return-ees from other destina-tions	Potential migrants
Russia	9.0	12.5	7.3	18.0	12.0	12.5	0.7	2.0
USA	15.0	6.1	15.0	10.0	9.5	3.0	11.3	23.0
France		7.0	8.7	8.0		3.5	4.3	10.0
Germany	6.5	1.0	3.3	4.0	4.0	3.0	7.7	3.0
Greece		0.5	2.3	5.0		-	0.3	-
Italy		4.0	3.3	9.0		3.0	4.0	11.0
Spain		1.5	2.3	6.0		3.0	5.0	5.0
Nether-lands		-	1.7	4.0		-	0.3	-
Great Brit-ain	0.5	1.0	1.7	3.0	1.5	3.5	6.7	11.0

12 Moscow - 'Extended homeland'?

Armenians and Georgians are no newcomers to the Russian capital and megalopolis of Moscow, as several Muscovite street names – e.g. 'Armyanskiy pereulok', 'Bol'šaya Gruzinskaya ulitsa' – and architectural heritage (churches, palaces of merchants) testify. The presence of Christian South Caucasians in Moscow dates back to the late 14th century when a permanent Armenian merchant colony emerged, specialized on the international trade with silk, textiles and oriental spices, and prosperous due to privileges granted by the Tsars. Several Armenian merchants and military commanders were raised to Russian peerage. Some of them became prominent sponsors of sciences and arts, such as the Lazarian (Russified: Lazarev) family that established in 1815 an institute for oriental and Caucasian languages (including Georgian), cultures and history (*Lazarev Institute for Oriental Languages*), whose Empire building is listed as a memorial building and currently houses the Embassy of the Republic of Armenia (Hofmann 2007: 383). Throughout the 19th and early 20th centuries, the *Lazarev Institute* and the adjacent Armenian print shop had been the main Armenian cultural center of Moscow.

The history of Georgians in Moscow reflects Georgian-Russian state interrelations: In the first half of the 18th century, the Georgian King Vakhtang VI., together with several noblemen immigrated to Moscow and built a palace there. Vakhtang's son Prince Giorgi, who was also a general of the Russian army, erected the still existing Church of St George on the same territory of 'Gruzino' in a blend of Georgian and Russian Orthodox styles. In the late 18th century, the Eastern Georgian King Erekle II sought the protection of the Russian Empire against Iranian invasions and overrule, safeguarded in the Treaty of Georgievsk (1783), which made the Kingdom of Kakheti and Kartli a Russian protectorate. But when Iran's new ruler Agha Mohammed Khan invaded and ravaged Georgia in 1785, Russia hesitated to help until 1876. Although the Treaty of

Georgievsk recognized Georgia's territorial integrity and the ruling dynasty of Bagrationi, Russia annexed Georgia's Eastern and Central provinces of Kakheti and Kartli and in 1810 the Western Georgian kingdom of Imereti.

Russian post-Soviet immigration policies showed little or no awareness of the intensity and peculiarities of Russian and South Caucasian relations. Armenians, Georgians, Azeris and immigrants that belonged to non-titular nations from the South Caucasus were increasingly treated as aliens and were at the same time collectively labeled as 'Caucasians', equally as the people of the North Caucasus, whose 'othering' in Russian society occurred as a result of Russia's conflict and wars against Chechnya.

In general, Russian immigration policies had been very uneven over the last two decades. In the early 1990es, when many refugees and displaced persons arrived from the previous Soviet republics, Russian immigration policies had been extremely permissive or even non existent. Among the refugees and DPs of those years were ethnic Russians as well as members of Russophone minorities, uprooted by the new linguistic nationalism in the independent states. A radical change took place, when in 2002 restrictive immigration laws and decrees against foreign residency and labor were issued, aiming at the regularization of immigration and resulting in the reduction of irregular immigrants from 50 to 15 percent.

The amendment of the RF immigration legislation in 2007, which led to partial re-liberalization, was opposed by administrators, including the then influential Mayor of Moscow (1992-2010), Yuri Luzhkov. (T)his opposition caused a certain restoration of the 2002 restrictions, exacerbated by the effects of the financial crisis. Luzhkov's successor Sergey Sobyanin is pursuing a similar protectionist policy, when making clear his populist opinion that Muscovites should be prioritized for Moscow jobs. Year-on-year the quota for foreign workers in Russia's capital is downsized by 70,000-to-100,000 people, egged on by ultranationalist, right-wing organizations who have clearly ascertained their platform of preserving a 'Russia for Russians' (Mack 2010).

Already in 2008, migrants were excluded from certain traditional for intra-CIS migration sectors of the economy, such as retail trade. The introduction of an electronic Employment authorization card in 2007, issued by the *Federal Migration Service*, freed immigrants to Russia from their previous dependency on their employers. But employers frequently fail to register their foreign employees at the FMS. The legal or illegal status of a foreigner in Russia therefore depends not at least on his or her employer[21] who might prefer to operate in the shadow economy in order to cut wages and save taxes in order to become more

21 In 2007, 40% of the authorized to work migrants were nevertheless unofficially hired: "A perfectly legal migrant may still turn out to be an illegal worker and may even not be aware of it" (Ioffe & Zayonchkovskaya 2010: 24).

competitive. The high share of Russia's shadow economy, estimated by the head of the RF Federal Statistics Service, Alexander Surinov to be as high as one fifth of the GDP in early 2010 (Abelsky 2010), indicates a likewise high percentage of undocumented labor in the country. Although the quota for Moscow allowed 810,000 immigrants from CIS states in 2007, their estimated real total was 953,000. In 2010, the official immigration quota for the Russian Federation was reduced to two million, including a 'reserve' of 0.7 million. In conclusion we may portrait Russian immigration policies as ambiguous, because divergent and even conflictual aims are pursued: RF immigration policies have to consider Russia's need for cheap unskilled labor and the low wages in the RF industry, construction, and forestry. 80% of all labor migrants in Russia are not required to have qualifications. There is also a strong demographic need for immigration and a subsequent liberal immigration policy. According to a UN projection, the population of the Russian Federation could decline from 142,499 (2007) to 107,832 million in 2050 (UN 2007: 65, Table A 2). With an annual shrinking of its population for 700,000 persons, Russia will need by 2050 a net immigrant population of at least 35.8 million to compensate the decline of its working-age population.

Current RF migration laws and policies have presented significant bureaucratic obstacles for foreigners hoping to live and work in Russia, in particular for non-CIS nationals. President Dmitry Medvedev has put the development of Russia's budding hi-technology sectors on the top of his priority list. Part of this process includes easing Russia's often cumbersome migration laws for highly-skilled professionals. A new bill being considered in the State Duma will allow foreign specialists to receive a renewable three-year work permit, a residency permit, and pay the same highly attractive 13% tax rate as legitimate Russian or CIS citizens (down from the current 30% for non-CIS foreigners). "While the spotlight focuses primarily on the competition for the world's top innovative talent, President Medvedev has a much broader vision in mind, hoping to attract 20 million new foreign workers over the next 10-to-15 years" (Mack 2010).

In Moscow, 84.5% of the more than 10 million residents are currently ethnic Russians. But according to a report created by the *City Committee on Interregional relations and National Policies* this situation will sharply change by 2025, when the working population of the city will decrease, significantly changing the dynamics of the metropolitan labor market. A growing percentage of the working population will then be ethnically non-Russian. The report further mentions that in the last decade, neighborhoods with concentrations of ethnic minorities began appearing. Recent migrants, particularly the Chinese, Vietnamese and "those from the Caucasus", tend to form a closed, isolated subculture, which in turn leads many Muscovites to feel that the culture of their city is being threatened by these newcomers. The report further says that eradicating extremism, xenophobia and domestic nationalism is a priority for the city authorities, and that improving migration legislation is essential. The city aims to crack down on the low-wage paying black market of labor, reduce illegal migration and work out a more "effective use of the foreign and regional workforce" (Houdouchi 2010).

Although experts and decision makers in Russia well understand the inevitability of foreign labor immigration, they fear the lacking societal acceptance in the indigenous population and also insoluble problems of adaption in the immigrant communities. Their fear for a change of the ethnic composition because of non-Slavic immigration, followed by increased xenophobia led to an attempted, but abortive ethnization of immigration and to insufficient projects to that aim[22]: According to the FMS, 20,000 voluntary Russian 'repatriates' have received papers to resettle in Russia; of these, only 13,800 actually did resettle (Adelaja 2010).

"Migrantophobia" has become part of the increasing societal xenophobia that has been observed in Russia since the 2000s. The xenophobia in turn appears as a result of racism that is discriminatory both against foreigners and RF nationals of non-Slavic origin. International bodies of the United Nations as well as RF human and civic rights NGOs have expressed profound concern about grave and wide spread violations of the rights of migrants residing in Russia and in particular in Moscow. The 2009 'Country Report on Human Rights Practices' (US Department 2010) of the *US Department of State* enlists the following violations of fundamental rights:

- **Right to work, freedom of movement**: Authorities frequently denied migrants the right to work if they did not have residential registration.
- **Children's right for education**: Although education was free to grade 11 and compulsory until age 15 or 16, regional authorities frequently denied school access to the children of persons not registered as residents of the locality, including Roma, asylum seekers, and migrants.
- **Forced labor (a crime against humanity), labor rights**: In a study of migrant construction workers published in February, the New York based *Human Rights Watch* (HRW) documented numerous cases of forced labor (Buchanan 2009), despite passage of the 2007 migration law requiring workers to register directly with the state. In all cases employers confiscated the workers' passports to coerce and confine them. Employers also

22 For the first time, an ethnisized approach towards immigration showed in the priority given to the repatriation of Russian expat communities. In his state-of-the-nation address of May 2006, the then President Vladimir Putin vowed to make Russia's population decline his highest priority and initiated changes in the migration policy, such as the attempt to attract compatriots from abroad. Subsequently, a six-year government programme was launched in June 2007, encouraging 'compatriots' who live abroad to return to Russia. They will receive cash, social benefits and support to gain or regain Russian citizenship. However, the migration potential of this group accounts for only 6-7 million people and is expensive: The reintegration of one million 'repatriates' will cost approximately US $ 6 billion (Krehm 2006).

withheld their wages, physically abused them, threatened to denounce them to the authorities, and induced indebtedness, which they then had to work to repay. HRW also documented incidents of police extorting and beating migrant workers and asserted that the executive or judicial agencies charged with addressing labor rights violations failed to investigate and ensure prosecution of violations effectively.

- **Labor Trafficking:** The *International Labor Organization* reported that labor trafficking was the predominant form of trafficking. In addition to the above mentioned restrictions, labor traffickers controlled their victims by such means as the use and threat of force, threats to report them to authorities, and confiscating their travel or personal identity documents.
- **Discrimination based on nationality/ethnicity**: The RF law prohibits discrimination based on nationality/ethnicity; however, government officials at times subjected minorities to discrimination. Recent years have also seen a steady rise in societal violence and discrimination against minorities, particularly Roma, persons from the Caucasus and Central Asia, dark-skinned persons, and foreigners. Although the number of reported hate crimes decreased during 2009, skinhead groups and other extreme nationalist organizations fomented racially motivated violence. In 2010, 37 persons died as victims of racial violence against ethnic Non-Russians (Schepp 2010: 92). Racist propaganda remained a problem throughout 2009 and 2010, although courts continued to convict individuals of inciting ethnic hatred by means of propaganda.
- **Insufficiencies in law enforcement and policing**: The 2009 *Country Report on Human Rights Practices* noted numerous and diversified problems of law enforcement in Russia and Moscow in particular. As in previous years, federal and local law enforcement personnel continued to target members of ethnic minorities disproportionately. Police reportedly beat, harassed, and demanded bribes from persons with dark skin or who appeared to be from the Caucasus region, Central Asia, or Africa. In Moscow authorities continued to subject dark-skinned persons to far more frequent document checks than others and frequently detained or imposed illegally large penalties on them for lacking documents. In a January 2009 raid, police confiscated presents that migrants had bought as New Year's gifts for family members back home. During one of several raids on the Čelobitevo shantytown, where many migrant laborers were housed, police officers reportedly forced a dozen men to strip to their underwear in frigid temperatures when they could not produce a sufficiently large cash bribe.

Police often failed to record the abuse of minorities or to issue written citations to the alleged victims. Law enforcement authorities frequently targeted such persons for deportation from urban centers. Police investigation of cases that appeared to be racially or ethnically motivated was often ineffective. Authorities were at times reluctant to acknowledge the racial or nationalist element in the crimes, often belittling attacks and hate crimes as "hooliganism". Many victims met with police indifference, and immigrants and asylum seekers who lacked residence documents and were recognized by police often chose not to report attacks. According to the independent Moscow based Russian human rights *SOVA* Center (Moscow), willingness to validate crimes as hate crimes varied widely depending on the personal views of the local prosecutor. The center also noted that the number of hate crimes prosecuted in Moscow increased significantly after a new prosecutor took office in 2008.

As in previous years, skinhead violence against Asians and Caucasians continued to be a serious problem throughout 2009 and 2010. According to the RF Home Ministry, neofascist movements had approximately 15,000 to 20,000 members in 2009, more than 5,000 of whom were estimated to live in Moscow. However, the ministry stated that if the category were expanded to include 'extremist youth groups' in general, the number was closer to 200,000. In February 2009 the *Moscow Bureau for Human Rights* (MBHR), which is funded by the European Union, estimated that there were up to 70,000 skinhead and radical nationalist organizations operating in the country compared with a few thousand in the early 1990s. Skinhead groups were most numerous in Moscow, St. Petersburg, Nižniy Novgorod, Yaroslavl, and Voronež. The three most prominent ultra-nationalist groups - the *Great Russia* party[23] (Velikaya Rossiya), the *Slavic Union* (Slavyanskiy Soyuz) movement, and the *Movement against Illegal Immigration* (Dviženie protiv Nelegal'noy Migratsii - DPNI)–claimed, respectively, 80,000, 10,000, and 20,000 members (US Department 2010).

Perpetrators of hate crimes do not distinguish between immigrants from the South Caucasus and RF nationals from the North Caucasus. Russia's wars in Chechnya, the growing Islamization and radicalization there and in adjacent Ingushetia on the one hand, and suicide attacks by Northern Caucasians on public transport facilities and cultural sites in Moscow on the other hand added to the insecurity of Muscovites and visitors to the Russian capital, thus fuelling xenophobia among the Slavic population. After the mayor of the town of Khotkovo near Moscow publicly announced 'self-cleansing' in October 2010, right-wing extremists set fire to a hostel of migrants, while shops dismissed all their Non-Slavic employees (Schepp 2010: 92).

However, a closer look into the efforts of RF law enforcement bodies reveals also certain progress as a new trend. The *Sova Center* which is monitoring hate crimes since 2004 concludes that racism and xenophobia represent not only a threat for Russia's minorities, but also a destabilizing challenge to the state:

"2009 was a year of significant change in terms of activities of radical nationalists and efforts to counteract the manifestations of racism and xenophobia in the country. The main outcome of 2009 was a clear reduction in the number

---

23 The RF Registration Service rejected the party's official registration in 2007.

of victims of racist and neo-Nazi motivated violence for the first time in six years of observation conducted by SOVA Center. To some extent, credit should go to the law enforcement agencies who suppressed the largest and most aggressive ultra-right groups in the Moscow region in the second half of 2008 and in 2009. However, despite all efforts xenophobic violence remains alarming in its scope and extends over most of the Russian regions, affecting hundreds of people. 2009 saw an unprecedented growth of racist vandals' activity. Vandalism in 2009 was primarily ideological, rather than (anti-)religious in nature.

The ultra-right groups are actively and deliberately switching to anti-state terrorism. Their objectives are to destabilize the government, to increase public distrust of the government, and to paralyze civil society organizations working to counteract racism and xenophobia. Apologists of the ultra-right terror see their ultimate goal in provoking 'a nationalist revolution' and establishing a neo-Nazi regime in Russia" (Kozhevnikova 2010).

Assessing Russia's public discourse on immigration and related matters one cannot help but conclude that immigration policies have become increasingly influenced by foreign concepts that have little in common with the reality of regional migration in the post-Soviet space. Although the people involved in this migration are basically the same then they used to be in the 1990s and even before, the perception of migration and related matters has adapted to the migration and integration policy in other migratory systems, especially the North American and the European systems. As a result of this internationalization migrants – including those from the post-Soviet space – are perceived as culturally alien and incompatible with Russian traditions. More and more RF integration policies follow concepts, as discussed in the EU, without realizing that not the migrants from post-Soviet space have changed, but the Russian perception of them. By the way, the "Europeanization" of the Russian public discourse on migration and integration issues can also be understood as a result of the closer integration of Russia into the partnership policies of the European Union.

Whether migrants originate from China, Vietnam, the post-Soviet Central Asia or the South Caucasus, in Russia they are indiscriminately seen as aliens and a threat to Russian culture and identity. Indicative for this trend, as well as for the above mentioned Europeanization is the attention given by RF mass media to the coverage of youth riots in the immigrant enclaves in the suburbs of French cities in November 2005. Russian media widely covered these events as "migrants' danger", which at the same time was instrumentalized by the political parties for attracting the attention of voters. It is against this back-ground of xenophobic populism that slogans like 'Russia for Russians' to greater or lesser extents are supported by two thirds of the Russian population (Charny 2005: 1).

Among many other results have the 2009 ArGeMi polls revealed that the respondents from Armenia and Georgia who reside in Moscow perceive their situation contrary to the average ethnic Russian Muscovite. These cohorts are highly adapted and integrated into the Muscovite/Russian society. 98.5% of the respondents from Armenia and 96% of those from Georgia assessed their Russian language skills as fluent or good. More than every second respondent holds the RF citizenship. A quarter of the respondents from Georgia and even a third of those from Armenia are married to ethnic Russians. While Russian integration experts believe this to be a rather low share, a comparison with Germany allows the opposite conclusion: Only three percent of the male migrants and eight percent of the young women from Turkey married a German partner, while in the second generation the ratio of inter-ethnic marriages increased only slightly from three to 8.9% among the males (Bode et al. 2010: 27). The ArGeMi polls on media consume, behavior in leisure time, frequency of links with the homeland etc., are further indicators for the high adaption, if not assimilation of Christian South Caucasians into the Muscovite society.

Contrary to the perception of migrants as threatening aliens the Muscovite residents from Armenia and Georgia perceive Russia as an 'extended homeland' that is familiar and dear to Armenians and Georgians since the late middle Ages. "Next to Constantinople and Tiflis, Moscow has played an important role in Armenian cultural and commercial history", mentioned an Armenian in private conversation, who does business in Moscow and Minsk, editing at the same time an Armenian journal in Russian language. "For Armenians, Moscow is not a far abroad. It is an extended homeland."

However, the repercussions and disappointments caused by Georgia's deteriorated relations with Russia resulted in re-evaluation. As the analysis of samples from Georgia reveals, the ranking of Moscow in the destination hierarchy is changing. Migrants from that country are well on their way to new and safer destinations in Europe and in the USA.

References

ABAZOV, Rafis (2009): *Current Trends in Migration in the Common Wealth of Independent States*. UNDP, Human Development Reports, Research Paper No. 36. - http://hdr.undp.org/en/reports/global/hdr2009/papers/HDRP_2009_36.pdf

ABDURAZAKOVA, Dono: *Social impact of international migration and remittances in Central Asia (with focus on Armenia, Azerbaijan, Kyrgyzstan, Tajikistan and Uzbekistan)*. United Nations Economic and Social Commission for Asia and the Pacific, Expert Group Meeting on Strengthening Capacities for Migration Management in Central Asia, 20 and 21 September 2010. - Bangkokhttp://www.unescap.org/sdd/meetings/egm_mig_sep2010/mig_egm_paper_dono.pdf

ABELSKY, Paul (2010): 'Russian 'Shadow' Economy Accounts for 20% of GDP.' *Bloomberg Businessweek*, 15 January 2010 http://www.businessweek.com/news/2010-01-15/russian-shadow-economy-accounts-for-20-of-gdp-update1-.html

ADELAJA, Tai (2010): 'Voluntary repatriation.' *Russia Profile.Org*, 15 July 2010 http://www.russiaprofile.org/page.php?pageid=Politics&articleid=a1279218307&print=yes

BARSOUMIAN, Nanore (2010): 'Poll Finds 39% of Armenians Wish to Leave Armenia Permanently.' *The Armenian Weekly*, 10 August 2010. - http://www.armenianweekly.com/2010/08/10/poll-finds-39-of-armenians-wish-to-leave-armenia-permanently/

BENDIXEN Sergio & Associates (2010): *Financial Sector Analysis for South Caucasus* - http://www.ebrd.com/downloads/sector/etc/finsec.pdf

BENDIXEN, Sergio & Associates (2007): *Summary of Findings from National Surveys: Azerbaijan, Georgia, Moldova.* - http://www.ebrd.com/downloads/sector/etc/natsum.pdf

BENDIXEN, Sergio & Associates (2007a): *Summary of Findings from Russia Survey: Russian Survey of Immigrants from Moldova, Georgia and Azerbaijan.* (June, 2007), http://www.ebrd.com/downloads/sector/etc/sumrus.pdf

BENDIXEN, Sergio & Associates (2007b): *Russian Survey of immigrants from Moldova, Georgia, and Azerbaijan. Survey presentation,* June http://www.ebrd.com/downloads/sector/etc/surru.pdf

BENDIXEN, Sergio & Associates (2007c): *Georgia National Public Opinion: Survey on Remittances.* Survey Presentation, July http://www.ebrd.com/downloads/sector/etc/surge.pdf

BODE, Kim et. al. (2010): „'Es gibt viele Sarrazins'". *Der Spiegel*, Nr. 36, 6 September 2010

bw-i (Baden-Württemberg International): Armenien – Daten und Fakten 2009 - http://www.bwglobal.de/deu/data/LaenderanalyseArmenien.pdf

BUCHANAN, Jane (2010): *"Are you happy to cheat us?": Exploitation of Migrant Construction Workers in Russia.* New York: Human Rights Watch, February 10, 2009. http://www.hrw.org/en/reports/2009/02/09/are-you-happy-cheat-us-0

CHARNY, Semyon (2005): *Xenophobia, migrant-phobia and radical nationalism at the elections to the Moscow City Duma; Review of the Moscow Bureau for Human Rights*. Moscow - http://antirasizm.ru/lv/english.php

CORSO, Molly (2011): 'Georgia: Evicted IDPs Enduring Winter without Heat, Water or Power'. *Eurasianet.org*, 31 January 2011. http://www.eurasianet.org/node/62800

GEVORKYAN, Aleksandr V., Arkady GEVORKYAN and Karine MASHURYAN (2008): *Little Job Growth Makes Labor Migration and Remittances the Norm in post-Soviet Armenia.* "Migration Information Source", March 2008 - http://www.migrationinformation.org/Feature/display.cfm?id=676

GINDLING, Tim H.; POGGIO, Sara: 'Family Separation and the Education Success of Immigrant Children.' *UMBC Policy Brief* No. 7, March 2009

GRADIROVSKI, Sergei; ESIPOVA, Nelli (2007): 'Conflict in the Caucasus; New Surveys on Azerbaijan-Armenia.' *Gallup Exclusive*, December 10, 2007 http://ipm.ge/article/GALLUP%20EXCLUSIVE-%20Conflict%20in%20the%20Caucasus%20ENG.pdf

GVALIA, Eka: 'Eviction or Integration?' *Eurasia Review*, 6 February 2011. - http://www.eurasiareview.com/analysis/georgia-evictions-or-integration-06022011/

HOFMANN, Tessa (2007): 'Armenische Handelskolonien in Russland seit dem Spätmittelalter' [Armenian Trade Colonies in Russia since the Late Middle Ages]. In: Klaus J. Bade et al. (Ed.): *Enzyklopädie Migration in Europa: Vom 17. Jahrhundert bis zur Gegenwart.* Paderborn, München, Wien, Zürich: Ferdinand Schöningh; Wilhelm Fink, pp. 382-385 (English edition: Cambridge University Press, Cambridge/New York [2011])

HOUDOUCHI, Ayano (2010): 'Migration could solve Moscow's problems.' *The Moscow News*, 23 February 2010. http://themoscownews.com/news/20100223/55414931.html?referfr

International Center for Human Development (2008): For Whom the Bell Tolls? The impact of the world financial crisis on migration in Armenia. Yerevan, 8 October 2008 http://www.ichd.org/?laid=1&com=module&module=static&id=367

IVAKHNYUK Irina (2009): 'Crises-related redirections of migration flows: The case of the Eurasian migration system. Paper presented to the Annual Conference "New Times? Economic Crisis, geo-political transformation and the emergent migration order".' University of Oxford http://www.compas.ox.ac.uk/fileadmin/files/pdfs/Non_WP_pdfs/Events_2009/Annual_Conference/Irina%20Ivakhnyuk%20-%20Crises-related%20redirections%20of%20migration%20flows.pdf

IVAKHNYUK Irina (2006): *Migration in the CIS region: Common problems and mutual benefits;* Paper presented to the International Symposium on International Migration and Development; Population Division, Department for Economic and Social Affairs, United Nations Secretariat, Turin, 28-30 June 2006

JOBELIUS, Matthias (2010): *Wirtschaftsliberalismus in Georgien: Verspielt die Regierung durch eine marktradikale Politik den Anschluss an Europa?* [Economic Liberalism in Georgia: Does the Government risk the admission to Europe through its market radical policy?], "Friedrich Ebert Stiftung", November 2010

JOHANSSON, Alice (2008): *Return Migration to Armenia: Monitoring the Embeddedness of Returnees.* Amsterdam; Nijmegen: Universiteit van Amsterdam; Radboud Universiteit Nijmegen, (January 2008) http://www.ru.nl/cidin/outreach_activities/research_projects/cordaid/

IOFFE, Grigory; ZAYONCHKOVSKAYA, Zhanna (2010): *Immigration to Russia: Why it is inevitable, and how large it may have to be to provide the workforce Russia needs.* Seattle: University of Washington, The National Council for Eurasian and East European Research, January 21, 2010. – http://www.ucis.pitt.edu/nceeer/2010_824-05g_Ioffe.pdf

ISHKANIAN, Armine (2003): 'Gendered transitions: the impact of the post-Soviet transition on women in Central Asia and the Caucasus.' *Perspectives on global development and technology*, 2 (3-4), pp. 475-496

ISHKANIAN, Armine (2002): 'Mobile motherhood: Armenian women's labor migration in the post-Soviet period.' *Diaspora: a journal of transnational studies*, 11 (3), pp. 383-415

JONES, Adele, SHARPE, Jacqueline and SOGREN, Michele (2004): 'Children's Experiences of Separation from Parents as a Consequence of Migration.' *Caribbean Journal of Social Work*, 3 (1), pp. 89-109

KARIMOV, Daniyar (2008): 'Kyrgyzstan's average wage rates second to last in CIS', 31 July 2008. - http://eng.24.kg/cis/2008/07/31/5732.html

KOZHEVNIKOVA, Galina (2010): *Under the Sign of Political Terror: Radical Nationalism and Efforts to Counteract It in 2009.* 10 March 2010. http://www.sova-center.ru/en/xenophobia/reports-analyses/2010/03/d18151/

KREHM, William (2006): 'Russia's population shrinks.' *Economic Reform*, May 2006 - http://www.sustecweb.co.uk/past/sustec14-3/russia.htm

LIVNY, Eric, OTT, Mack and TOROSYAN, Karine (2007): *Impact of Russian Sanctions on the Georgian economy.* Tbilisi: International School of Economics at Tbilisi State University www.iset.ge/.../impact_of_russian_sanctions_on_the_georgian_economy_-_final_version.ppt; http://www.iset.ge/files/conclusions_handout-revised.pdf

LUNDKVIST-HOUNDOUMADI, Margharita (2010): 'Treading on the fine line between self-sacrifice and immorality: Narratives of emigrated Georgian women.' *Transcience Journal*, Vol. 1, No. 2, pp. 50-70. - http://www.transcience-journal.uni-freiburg.de/Issue%202/Vol1_Issue2_2010_50_70.pdf

MACK, David (2010): *Russia Stands in its Own Way on Migration Reforms.* Center for Strategic and International Studies, December 14, 2010. – http://csis.org/blog/russia-stands-its-own-way-migration-reform

MANSOOR Ali, QUILLIN, Bryce (Eds.) (2007): *Migration and Remittances: Eastern Europe and the Former Soviet Union.* Washington: The International Bank for Reconstruction and Development; The World Bank

MARTIROSYAN, Narek Armen (2008): *Armenia's Current Practice Of Targeting Social Assistance And Analysis of Development Prospects.* Yerevan: USAID Armenia, SPSS - http://www.spss.am/data.php/325.pdf

MINASYAN Anna, POGHOSYAN, Alina; HAKOBYAN, Tereza and HANCILOVA, Blanka (2007): *Labor Migration from Armenia in 2005-2007: A Survey.* Yerevan: Asoghik http://www.osce.org/publications/oy/2007/11/28396_996_en.pdf

MINASYAN Anna; HANCILOVA, Blanka (2006): *Labor Migration from Armenia 2002-2005: A Sociological Survey of Households.* Yerevan: OSCE; Advanced Social Technologies, February 2006 http://www.osce.org/publications/oy/2006/02/18193_531_en.pdf

NARODONASELENIe [Population] (1994). Moscow: BRE

POMFRET Richard (1995): *The Economies of Central Asia.* Princeton: Princeton University Press

POTTINGER, Audrey M.; WILLIAMS BROWN, Sharon (2006): *Understanding the Impact of Parental Migration on Children: Implications for Counseling Families from the Caribbean.* - http://counselingoutfitters.com/Pottinger.htm

ROBERT, Bryan W.; BANAIAN, King (2004): *Remittances in Armenia: Size, Impacts, and Measures to Enhance their Contribution to Development.* Yerevan: USAID/Armenia, 1 October 2004

SADOVSKAYA Elena Y. (2006): 'International Labor Migration and Remittances in Central AsianRepublics: Strategy for Survival or Development?' In: International Migration: Proceedings of Moscow State University, Vol. 18. Moscow: TEIS (in Russian and English), pp. 38-36

SCHEPP, Matthias (2011): 'Russia: Ein Bazillus namens Hass' (A Virus named Hate). *Der Spiegel*, No. 5, 31 January 2011

SELM, Joanne van (2005): *Georgia Looks West, But Faces Migration Challenges at Home.* "Migration Information Source", June 2005 http://www.migrationinformation.org/Profiles/display.cfm?ID=314

'Social Housing Program Launched in Armenia' (2010). *ArmeniaNow.Com*, 03 April 2010. - http://www.armenianow.com/social/22259/social_housing_program_launched_in_armenia

SUAREZ-OROZCO Carola; Irina TODOROVA and J. LOUIE: 'Making up for lost time: The experience of separation and reunification among immigrant families.' *Family Process*, 2002, 41, pp. 625-643

TUMASYAN Mushegh; MANUKYAN, Yelena; TOROSYAN, Gagik; TERZIKYAN, Genady and MNATSAKANYAN, Varsenik (2008): *A Study on International Migrants' Remittances in Central Asia and South Caucasus; Country Report in*

Remittances of International Migrants and Poverty in Armenia. Manila: Asian Development Bank, December 2008. - http://www.adb.org/Documents/Reports/Consultant/40038-REG/40038-05-REG-TACR.pdf

(UN 2010) United Nations, Department for Economic and Social Affairs, Population Division (2007): *World Population Prospects: The 2006 Revision; Highlights.* New York http://www.un.org/esa/population/publications/wpp2006/WPP2006_Highlights_rev.pdf

(UNDAF 2010) *United Nations Development Assistance Framework 2010-2015 Armenia* http://planipolis.iiep.unesco.org/upload/Armenia/Armenia_UNDAF_2010-2015.pdf

(UNDP 2009) United Nations Development Program: *Human Development Report 2009; Overcoming barriers: Human mobility and development* http://hdr.undp.org/en/media/HDR_2009_EN_Complete.pdf

(UNDP 2009a) United Nations Development Program Armenia: *National Human Development Report 2009; Migration and Human Development: Opportunities and Challenges.* - http://hdr.undp.org/en/reports/national/europethecis/armenia/NHDR-2009-Armenia-EN.pdf

UNHCR (2010), Division of Programme Support and Management: *Asylum Levels and Trends in Industrialized Countries 2009: Statistical Overview of Asylum Applications Lodged in Europe and Selected Non-European Countries.* 23 March 2010 http://www.unhcr.ch/fileadmin/unhcr_data/pdfs/aktuell/AsylumReport2009.pdf

USAID Armenia, SPSS (2008): *Armenia's Labor Market Surveys* http://www.spss.am/data.php/341.pdf

U.S. Department of State (2010), Bureau of Democracy, Human Rights, and Labor: *2009 Human Rights Report: Russia.* Washington, March 11, 2010. - http://www.state.gov/g/drl/rls/hrrpt/2009/eur/136054.htm#

'WB: remittances from labor migrants to Armenia to reach $824mln in 2010.' (2010) *Panarmenian.net*, November 11, 2010. - http://www.panarmenian.net/eng/economy/news/56626/WB_remittances_from_labor_migrants_to_Armenia_to_reach_824mln_in_2010

World Bank (2011), International Bank für Reconstruction and Development: *Migration and Remittances Factbook 2011.* 2nd edition. Washington

Abbreviations

AMD	Armenian Dram (currency)
ArGeMi	Research project "Comparing Out-Migration from Armenia and Georgia"
ArmStat	State Statistical Office of the Republic of Armenia
ABD	Asian Bank of Development
CIS	Commonwealth of Independent States
DP(s)	Displaced person(s)
EBRD	European Bank for Reconstruction and Development
EUROSTAT	Statistical Office of the European Union
FMS	Federal Migration Service (of the Russian Federation)
GDP	Gross domestic product
HRW	Human Rights Watch
IDP(s)	Internally displaced person(s)
ILO	International Labor Organization
IMF	International Monetary Fund
IOM	International Organization for Migration
NGO	Non-Governmental Organization
NIS	Newly Independent States
OSCE	Organization for Security and Co-operation in Europe
RA	Republic of Armenia
RF	Russian Federation
RSFSR	Russian Socialist Federal Soviet Republic
RUB	Russian Ruble (currency)
UNDAF	United Nations Development Assistance Framework
UNDP	United Nations Development Program
UNHCR	United Nations High Commissioner for Refugees
USD	United States Dollar

Contributors

Badurashvili, Irina, Dr. sc., Tbilisi; sociologist; director of the *Georgian Center of Population Research (GCPR).* Research focus: Statistics; migration issues; social standard. Participant and head of many research projects in collaboration with or commissioned by foreign or international organizations such as the *European Union, Open Society Institute, World Bank, International Organization for Migration, Foundation for Population, Migration and Environment* (Switzerland).

Genov, Nikolai, Dr. sc.; Professor of Sociology at Freie Universität Berlin. Research fields: social theory, globalization, interethnic relations, social risks, international migration. Recent book publications: *Ethnicity and Mass Media* (2006), *Comparative Research in the Social Sciences* (2007), *Interethnic Integration* (2008), *Global Trends in Eastern Europe* (2010). Coordinator of research projects of EU, UNESCO, UNDP, ILO, Friedrich Ebert Foundation, Volkswagen Foundation. Recent comparative projects: *Comparing the Integration of Ethnic Minorities* (EU, 2005-2007), *Out-migration from Armenia and Georgia* (Volkswagen Foundation, 2008-2010). Director of the UNESCO-ISSC International Summer School *Comparative Research in the Social Sciences.*

Osadchaya, Galina I., PhD., Professor; Moscow; Sociologist; Vice-Rector of *Russian State Social University*; Research focus: social policy; migration policy; standard of living; family studies; member of the *Russian Academy of Social Sciences*; member of ISA, ESA; vice-president of the *Union of Sociologists of Russia*; vice-president of the *Russian Society of Sociologists.*

Poghosyan, Gevork, Dr. sociol., Professor, Corresponding member of *Armenian National Academy of Sciences* (ANAS); Yerevan; Sociologist; Director of the *Institute of Philosophy, Sociology and Law* of ANAS; President of the *Armenian Sociological Association*; Research focus: transition, migration, political processes in modern Armenia and social changes.

Savvidis, Tessa, Dr. phil., M.A.; Berlin; Philologist (Slavic literatures and languages; Armenian studies) and sociologist; Research Associate at the Chair of Eastern European Sociology (Institute for Eastern European Studies, Freie Universität Berlin); Research focus: migration and diaspora studies; ethnic minorities and conflicts in the South Caucasus and Middle East; author and editor of numerous monographs on the history, culture and present situation of Armenia.

Peter Lang · Internationaler Verlag der Wissenschaften

Alpago Alpago

Immigration – Isolation – Integration

Is that the only solution?

Frankfurt am Main, Berlin, Bern, Bruxelles, New York, Oxford, Wien, 2011.
175 pp., num. fig., tab. and graph.
ISBN 978-3-631-61440-2 · hb. € 31,80*

The target of this project is to consider cross-border mobility of economic, cultural and historical happenings from a long-term historical perspective. And moreover to deal with humanitarian issues such as peace, integration, immigration and peacefully living together as well as providing long-term perspectives for changes and promotion of peaceful change in society. The study breaks down the barriers and leads the readers to go beyond the borders. The components of the project are all focused on the sociological issues like immigration, isolation and integration. The method, which has been used, is a new kind of way to investigate sociological phenomena and interpret them in a manner which provides the background of challenges in a concrete and uncomplicated way with emotion and reality. The theoretic motto is: "Science for all". The project is particularly applicable for anthropologists, ethnologists, sociologists and philologists.

Content: The Dreams and Nightmares of a Refugee Teacher in Europe · Immigration, Isolation, Integration in Empirical Facts and Statistics · Demographic Challenges of the European Union in Facts and Statistics · Immigration in the USA in Facts and Statistics · Poverty and Immigration in Facts and Statistics · Visual Field Study

Frankfurt am Main · Berlin · Bern · Bruxelles · New York · Oxford · Wien
Distribution: Verlag Peter Lang AG
Moosstr. 1, CH-2542 Pieterlen
Telefax 00 41 (0) 32 / 376 17 27

*The €-price includes German tax rate
Prices are subject to change without notice

Homepage http://www.peterlang.de